URBAN ECONOMIC CHANGE

Publications in the Inner Cities Research Programme Series

1 Overview volume
Urban Economic Adjustment and the Future of British Cities: Directions for Urban Policy
Victor A. Hausner, Policy Studies Institute, London

2 Summary volume on five city studies
Urban Economic Change: Five City Studies
Members of the five research teams and the National Director

3 Bristol region study
Sunbelt City? A Study of Economic Change in Britain's M4 Growth Corridor
Martin Boddy, John Lovering, and Keith Bassett

4 Clydeside conurbation study
The City in Transition: Policies and Agencies for the Economic Regeneration of Clydeside
edited by W. F. Lever and Chris Moore

5 London metropolitan region study
The London Employment Problem
Nick Buck, Ian Gordon, and Ken Young, with John Ermisch and Liz Mills

6 West Midlands study
Crisis in the Industrial Heartland: A Study of the West Midlands
Ken Spencer, Andy Taylor, Barbara Smith, John Mawson, Norman Flynn, and Richard Batley

7 Newcastle metropolitan region study
Economic Development Policies: An Evaluative Study of the Newcastle Metropolitan Region
Fred Robinson, John Goddard and Colin Wren

8 Special research studies
Critical Issues in Urban Economic Development, vol. I
edited by Victor A. Hausner, Policy Studies Institute, London

9 Special research studies
Critical Issues in Urban Economic Development, vol. II
edited by Victor A. Hausner, Policy Studies Institute, London

10 Special research study
The New Economic Roles of UK Cities
Iain Begg and Barry Moore, Department of Applied Economics, University of Cambridge

11 Collective volume
Economic Change in British Cities
Members of the five research teams and the National Director

URBAN ECONOMIC CHANGE

Five City Studies

VICTOR A. HAUSNER
and members of the
ESRC INNER CITIES RESEARCH PROGRAMME

CLARENDON PRESS · OXFORD
1987

Oxford University Press, Walton Street, Oxford OX2 6DP
Oxford New York Toronto
Delhi Bombay Calcutta Madras Karachi
Petaling Jaya Singapore Hong Kong Tokyo
Nairobi Dar es Salaam Cape Town
Melbourne Auckland
and associated companies in
Beirut Berlin Ibadan Nicosia

Oxford is a trade mark of Oxford University Press

Published in the United States
by Oxford University Press, New York

British Library Cataloguing in Publication Data
Hausner, Victor
Urban economic change: five city studies.
—(Inner cities research programme; 2).
1. Urban economics 2. Cities and towns—Great Britain 3. Great Britain—Economic conditions—1945–
I. Title II. Inner Cities Research Programme III. Series
330.941'085'8 HC256.6
ISBN 0–19–823280–2

Library of Congress Cataloging in Publication Data
Urban economic change.
Publications in the inner cities research programme series: 2)
Includes index.
1. Urban economics—Case studies. 2. Urban policy—Great Britain—Case studies. I. Hausner, Victor A.
II. Inner Cities Research Programme (Great Britain)
III. Series.
HT321.U2913 1987 330.9173'2 87–7775
ISBN 0–19–823280–2

Set by Promenade Graphics Ltd., Cheltenham
Printed and bound in Great Britain by
Biddles Ltd., Guildford and Kings Lynn

FOREWORD

In 1982 the Environment and Planning Committee of the Economic and Social Research Council (SSRC at the time) initiated a three year comparative research programme to examine inner city problems within the broad context of major structural and spatial changes occurring in Great Britain. The programme was developed by the then SSRC Inner Cities Working Party, chaired by Professor Peter Hall, and subsequently the Executive Panel on the Inner Cities, chaired by Professor Gordon Cameron. The proposal for the research was originally described in 'A Research Agenda' (chapter 8) of *The Inner City in Context* (ed. P. G. Hall, Heinemann, 1981). The purpose of the programme was to examine the processes of urban change, the effects on urban socio-economic welfare, and the prospects, constraints, and requirements for more successful urban adjustment to structural change. The programme arose from concerns with the urban problems of economic decline, labour market imbalances, social distress, and the effectiveness of public policies in addressing these problems. It was hoped that the programme's findings would be useful for the improvement of public policies to strengthen urban economies (that is, foster growth, employment, and competitiveness), alleviate the distress caused by change and improve the conditions of distressed inner city areas and deprived urban residents.

In practice, the overall programme focused on the economic aspects of urban change. Specifically, the programme sought to identify the key factors affecting urban economic change and to describe and explain the processes of local economic change. Secondly, it aimed to describe the consequences of change for the urban economy and the employment of its residents. Finally, the programme attempted to assess the effects of public policies on the process of change.

The programme was based on the idea that there is diversity in the economic performance of different urban and inner city areas in the UK and in their adjustment to external forces of change; for example, changes in business competition, technology, and residential patterns. A comparative examination of the nature, processes, and effects of economic change on different urban centres should help to clarify and explain the differences in the experiences of economic change among and within urban areas and to identify those factors (including public policies) which impede or facilitate urban and inner city adjustment to change: that is, economic growth, increased business competitiveness, employment generation, employment for the disadvantaged in urban labour markets, and the effectiveness of urban economic policies. The ESRC programme utilized a general framework of topics to assist individual urban studies and the comparative examination of urban economic change. The topics were: the changing nature of urban economic problems; the nature and causes of imbalances in urban labour markets; the unintended effects of central and local government policies on urban economic change, and the effectiveness of national and local urban economic policies; and the capabilities of local authorities to design and implement more effective economic development policies.

The Inner Cities Research Programme addressed these concerns through two avenues of work. The core of the programme was independent studies of four major urban centres of Great Britain which were selected by the SSRC as examples of the diversity of urban economic adjustment experiences: Glasgow and the Clydeside conurbation as an example of 'persistent economic decline'; Birmingham and the West Midlands conurbation as one of 'faltering growth'; the Bristol region as one of 'successful adaptation'; and two areas in London, one (Greenwich, Southwark, and Lewisham) exhibiting 'severe problems', the other (Brent, Ealing, and Hounslow) exemplifying more 'successful adaptation'. These were largely secondary research studies using existing data and analysing existing research. The studies were conducted by independent research teams with distinctive approaches and concerns, but linked by the overall programme's objectives and general framework of topical concerns, dialogue, information exchange, and some common data and analysis. The different research teams decided on the particular approach, subjects, and hypotheses which they considered most relevant to an understanding of economic change in their study areas; and critically examined the characterizations of those areas.

To these four initial studies a fifth study of the Newcastle metropolitan region was added, funded by the Department of the Environment, in order to broaden the sample of urban areas and focus particularly on an assessment of the relationship, impacts, and effectiveness of central and local government urban and regional economic development policies on an economically distressed city region that was a long-term recipient of government assistance.

In addition to these five core studies a number of smaller 'cross-cutting' studies were conducted by various researchers in order to provide a national statistical framework for the five city studies and uniform comparative data on the five city study areas, to examine in greater detail important aspects of urban economic change in a broader sample of urban areas, to explore the effects on change of important public policies, to provide a comparative international perspective, and to increase the general relevance of a case-study-based research programme. A list of the publications and their authors, resulting from all of these elements of the programme, appears at the beginning of this book.

In order to enhance the relevance of the research programme to public policy issues and communication with central government policy-makers, discussions were held with officials of three government agencies: the Department of the Environment, the Department of Trade and Industry, and the Manpower Services Commission. These agencies also assisted the programme through the provision of data, special analyses, and the co-operation of their regional offices. The individual city and cross-cutting studies involved extensive local contacts with government officials, representatives of business, labour and voluntary organizations, and other researchers and analysts.

The five city studies were conducted by members of research teams at the following institutions: the Department of Social and Economic Research at the University of Glasgow; the Institute of Local Government Studies and the Centre for Urban and Regional Studies at the University of Birmingham; the School for Advanced Urban Studies and the Department of Geography at the University of Bristol; the Policy Studies Institute in London and the Urban and Regional Studies Unit at the University of Kent; and the Centre for Urban and Regional Development Studies at the University of Newcastle upon Tyne.

Professor Brian Robson, chairman of the ESRC Environment and Planning Committee, and Professor Noel Boaden and Paul McQuail, both members of the Committee, advised on the implementation of the programme. They were assisted by Dr Angela Williams, Senior Scientific Officer to the Committee. Members of the former SSRC Executive Panel discussed, reviewed papers, and advised on the research during the course of the programme. The programme also benefited from the advice and comments of other urban analysts.

VICTOR A. HAUSNER
Director, ESRC Inner Cities Research Programme

London 1985

ACKNOWLEDGEMENTS

Thanks go to the Economic and Social Research Council for providing the financial support for the research which is reported in this volume. The Department of the Environment assisted also through the funding of the project on the Newcastle Metropolitan Region. The Policy Studies Institute (PSI) provided a supportive home from which the ESRC Inner Cities Research Programme was managed. Special thanks are due to Rosemary Lewin of PSI for invaluable administrative assistance during the course of the Research Programme and in preparing this and other volumes in the Programme series for publication.

CONTENTS

LIST OF FIGURES

LIST OF TABLES

1

Introduction: Economic Change and Urban Policy

*Victor A. Hausner**

The United Kingdom, like other Western advanced industrial nations, is in the grip of profound economic and social transformations. Adjustment to change—structural, technological, occupational, competitive, and demographic—is the order of the day. Key economic changes include alterations in the international division of labour, the decline of traditional manufacturing, increased labour-saving investment in new technology, reduced demand for unskilled labour and the rise of part-time female employment, and changes in the character of business competition. Slow economic growth, recession, and very high levels of national unemployment have been fostered by these structural changes. Changes in government economic and fiscal policies have also affected the adjustment process. The decentralization of population and employment from inner cities, larger urban areas, and older industrial regions and changes in urban economic functions have been other important elements of this process.

These changes have affected the national economy and the geographic and social distribution of economic activity, growth, and well-being. They have placed the economies and social orders of many of the nation's urban areas—particularly the older, major industrial cities—under serious pressure, instigating a process of disinvestment in industry and a broader process of economic, environmental, and residential decline. We are witnessing a process of urban restructuring linked to changes in advanced industrial societies affected by a changing international economic environment. Cities have experienced long-term trends of decline and increasing unemployment problems relative to the nation. However, there have also been variations in urban economic performance. In-depth studies of particular urban areas can both detail the processes of urban economic change and help to identify those factors which account for differences in performance between areas and influence the ability of an area to adapt to change.

Such a comparison of urban experiences can also contribute to a better understanding of the changing spatial economy of Britain, the new economic roles of cities, and the potential growth opportunities for urban economies. The industrial and occupational characteristics of the more successful urban

* Victor A. Hausner was, at the time of writing, the National Director of the ESRC Inner Cities Research Programme.

economies indicate the changes taking place in the hierarchy of the nation's urban areas.

Such studies should also provide guidance on the potential for intervention in urban economic problems and in designing area development policies which can contribute to long-term urban economic regeneration and a more equitable distribution of economic opportunity among the nation's population. In this critical transition period for cities, an improved understanding of the processes of economic change should help to foster development policies which can effectively promote, not the recapture of an urban economic past linked to earlier stages of industrial development, but the achievement of the best of the alternative economic futures open to cities and all their citizens.

The increase of national unemployment and industrial restructuring have expanded concern from isolated inner cities to the economic problems of entire urban areas and their regions, and have encouraged a recognition of the relationship between the plight of the inner cities and wider processes of change. The role of urban and regional redevelopment in national economic renewal has also become an important issue. Are sub-national development policies necessarily in conflict with national economic adjustment and revitalization, or can they make a useful, and possibly even an essential, contribution to national development? At the same time, the severe and disproportionate effects of increasing unemployment on inner city and other disadvantaged urban residents have shifted attention from social needs to employment needs and local job creation. These concerns have in turn affected urban policy, which has since the mid-1970s emphasized economic and property redevelopment, although still within inner city locations. Local authorities, with increased financing from the government's Urban Programme, have expanded their development and job creation activities. However, recent social conflict and violence in inner city areas have brought renewed attention to social problems and the need to 'target' policies on specific groups and communities within cities. The Economic and Social Research Council (ESRC) Inner Cities Research Programme addressed these dual economic concerns with the performance of the urban economy and the employment problems of disadvantaged urban residents, and assessed the relevance and adequacy of public policies to each.

The following essays report on five studies of a diverse group of large British urban areas which have had varying success in adjusting to change. They summarize many of the major findings of the city studies which were the core of the ESRC Inner Cities Research Programme. The individual city studies address the interaction of broad external trends with local conditions and trace the effects on the structure of the local economy, employment, the labour market, inner city areas, and disadvantaged urban residents. Most importantly, they attempt to identify those forces and public policies which account for the distinctive responses of the areas to change. What are the attributes of some urban areas which provide them with competitive advantages in the changing economic environment? Are there lessons from more successful urban economies that could be emulated by more troubled urban areas? What are the implications for policy-makers concerned with strengthening the urban region economy and providing employment for the unemployed and those disadvantaged in the competition for employment?

Several issues require attention in reviewing the findings of these studies: the indicators of area economic performance and economic welfare, the factors which affect the adjustment to change, and the effects of public policy on urban economic change. First, what are the indicators of (1) urban economic performance and (2) economic adjustment to change? Special studies in the ESRC Programme (Wolman 1985; Begg *et al.* 1986) focused on a few statistical indicators of economic performance—namely, employment growth (a measure of the ability of the urban economy to generate jobs) and unemployment (a measure of the ability of the urban economy to provide employment for its labour force). Other researchers have used a variety of quantitative indices of local economic performance including population growth and measures of income (Champion and Green 1985). However, the concept of economic adjustment (and development) extends beyond quantitative indices of growth to qualitative changes in area economies which underlie these statistics and their implications for local development. Is an area shifting into expanding sectors of economic activity and employment growth? Is the competitiveness and growth potential of its economic base improving? Are its physical, human, and financial resources being redeployed for productive use and to foster growth? In order to foster both growth and development, urban economic development policies must seek to reverse the decline of economic activity and employment, and facilitate the adjustment of the economic base to activities with growth potential.

Secondly, the economic welfare of urban residents as a whole and of specific subgroups (for example, ethnic minorities, women, and inner city residents) is also important in assessing the success of an area in adjusting to change. Indicators of economic welfare include: income, rates and length of unemployment, the quality of employment (for example, stability, wages, and potential for advancement), and the distribution of employment and income among different social groups. In the light of this distinction between urban economic performance and the economic welfare of urban residents, the relationship between the two needs to be examined. To what extent does local economic growth improve local economic welfare? Does local employment growth affect local unemployment rates? Do the benefits of local economic growth accrue to the disadvantaged members of the community?

Thus, one major concern of the ESRC Programme has been labour market adjustment problems in the changing urban economy under the constraints of high national and regional unemployment. Who are the disadvantaged in the urban labour market? How have they fared as a result of the changes in the urban economy, and in more and less successful urban economies? What are the major impediments to increasing their employment opportunities in the short and long term, and have public policies adequately addressed them?

Thirdly, it is important to identify not only the forces of change affecting urban areas but also those factors that impede or facilitate successful adjustment to change. What are the comparative advantages of the more successful urban economies? What are the locational requirements of the new growth sectors—for example, business services and advanced technology? How open to intervention are the conditions which influence the urban economic adjustment process?

Lastly, the effects of past and present public policies on the structure and competitiveness of urban economies and on the generation and distribution of employment must be assessed. This encompasses explicit area and industrial development and employment policies, and those other policies which have had significant if unintentional effects on urban economic change. Such information can assist in the formulation of more effective policies in the future.

The five city studies provide significant insights into these issues and offer some valuable recommendations regarding urban economic adjustment policy. In addition to reporting on some of the studies' findings, this overview essay places those findings in the context of national urban trends, introduces relevant findings from the programme's other research, and outlines some of the policy implications of the research.

Background: National Urban Trends, Problems, and Performance

The performance of the five city study areas must be assessed in the context of general urban trends and the performance of the British urban system.

There has been a major shift in the geographic distribution of population and industry in the post-war period (Begg *et al.* 1986). Population and employment have relentlessly deserted Britain's larger urban areas, particularly the inner cities of the six large conurbations,[1] while smaller towns and rural areas have experienced rapid growth. The working age population in Great Britain increased by 3 per cent between 1951 and 1981, while that of smaller towns and rural areas increased by 21 per cent; however, the six large conurbations as a whole experienced a 20 per cent decline, the inner areas a 33 per cent decline, and the seventeen other large free-standing cities[2] a loss of 6 per cent (Table 1.1). While national employment in these three decades increased by 3 per cent, total employment grew by 18 per cent in the smaller towns and rural areas, but fell by 12 per cent in the six large conurbations, and by 26 per cent in their inner areas (Table 1.2). Residential urban employment fell by even greater amounts, particularly in the inner cities of the six large conurbations, with a substantial 43 per cent decline. By 1981, almost 39 per cent of the reduced number of inner city jobs were taken by

Table 1.1. *Changes in Population of Working Age Groups in Areas of Great Britain, 1951–1981*

Type of area	1951	1961	1971	1981
Inner areas of six large conurbations	100	95	79	67
Outer areas of six large conurbations	100	94	90	90
Six large conurbations	100	94	85	80
Seventeen free-standing cities	100	98	97	94
Small towns and rural areas	100	108	117	121
Great Britain	100	102	103	103

Note: The figures show population of working-age groups (index, 1951 = 100).
Source: Adapted from Table 2.10 in Begg *et al.* (1986).

Table 1.2. *Changes in Total Employment in Areas of Great Britain, 1951–1981*

Type of area	1951	1961	1971	1981
Inner areas of six large conurbations	100	101	86	74
Outer areas of six large conurbations	100	107	109	99
Six large conburations	100	104	99	88
Seventeen free-standing cities	100	106	105	100
Smaller towns and rural areas	100	111	116	118
Great Britain	100	108	107	103

Note: The figures show total employment (index, 1951 = 100).
Source: Adapted from Table 2.9 in Begg *et al.* (1986).

inward commuters—nearly twice the proportion in 1951. Prior to the 1978 recession, the great majority of new jobs created in the 1970s were located in the smaller free-standing cities and towns. The conurbations lost the largest amounts of employment. In the first years of the recession, 1978–81, the conurbations and their surrounding metropolitan regions, with the exception of London, again suffered the greatest employment losses. (The 1980s have provided signs in London of a possible reversal of the urban–rural shift.) While the inner cities were hit disproportionately hard, there are indications of the spread of the inner city problem of relative employment decline to the outer areas of the conurbations. The inner cities of the conurbations continued to lose employment at twice the rate of the outer areas and free-standing cities, but their relative rate of decline slowed in the 1970s. At the same time, employment decline in the outer areas increased both relatively and absolutely, particularly in the 1970s. The large free-standing cities also experienced relative employment decline in the 1970s, but more modestly than the outer areas of the conurbations.

At the heart of urban employment loss has been the national decline in manufacturing employment (a loss of over 1 900 000 jobs between 1971 and 1981) which disproportionately affected the large urban areas where it was concentrated at the beginning of the post-war period. Between 1951 and 1981 the inner cities of the six large conurbations lost over one million manufacturing jobs—half the manufacturing jobs with which they began the period. The outer cities of the conurbations and other free-standing cities lost another million manufacturing jobs. In all three of these groups of urban areas, manufacturing job loss has consistently exceeded the average for Great Britain. In contrast to manufacturing, private and public services have been a major source of employment growth nationally. However, while service sector employment increased in the free-standing cities and in the outer cities of the conurbations during the 1960s and 1970s, employment in both these sectors declined in the inner cities of the six conurbations during the 1970s (Table 1.3).

Concomitant with population and employment decline has been an increase in the unemployment problems of urban residents both absolutely and relative to the nation. In 1981 the unemployment rate in the conurbations was 22 per cent above the national average; this compares with the rate for the other free-standing cities (15 per cent above the national average) and that for smaller towns and rural areas (10 per cent below the national average). The problem was particularly severe in the inner city areas of the conurbations. In 1981 residential unemployment in the inner cities of the six conurbations was 51 per cent above the national average, up from 33 per cent above average in 1951. However, unemployment problems were no longer confined to inner areas. Whereas, prior to 1981, residential unemployment in the outer areas of the conurbations was consistently below that of the nation, by 1981 these outer city areas also had unemployment rates which exceeded, albeit by only one percentage point, those of the nation (Table 1.4).

Closely related to these changes in the nation's population and industrial settlement pattern has been growing social polarization characterized by increasing concentrations of unemployed and deprived people in the inner cities of the conurbations. The inner cores of the conurbations dominate the list of the most deprived areas of the nation, and their severe deprivation problems are closely linked to population loss, particularly the relative out-migration of more affluent inner city residents (Table 1.5). Severe deprivation is now also to be found on the newer peripheral estates.

Urban unemployment and social problems are closely correlated with and exacerbated by the other growing polarization in Britain—that between the economically distressed, mainly northern, regions and the relatively prosperous south of England. With the exception of London, employment decline during the period 1971–8 was concentrated in the urbanized areas of the Midlands, northern England, South Wales, and central Scotland. Successful local employment growth was concentrated in the south of the country, and the regional bias against the north became much more marked in the recent

Table 1.3. *Changes in Employment by Sector and Type of Area in Great Britain, 1951–1981*

Sector and periods	Inner cities		Outer cities		Free-standing cities		Small towns and rural areas		Great Britain	
	(000s)	(%)	(000s)	(%)	(000s)	(%)	(000s)	(%)	(000s)	(%)
Manufacturing										
1951–61	−143	−8.0	+84	+5.0	−21	−2.0	+453	+14.0	+374	+5.0
1961–71	−428	−26.1	−217	−10.3	−93	−6.2	+489	+12.5	−255	−3.9
1971–81	−447	−36.8	−480	−32.6	−311	−28.6	−717	−17.2	−1929	−24.5
Private services										
1951–61	+192	+11.0	+110	+11.0	+128	+17.0	+514	+16.0	+944	+14.0
1961–71	−297	−15.3	+92	+8.1	−7	−0.8	+535	+14.5	+318	+4.2
1971–81	−105	−6.4	+170	+17.3	+91	+10.9	+805	+24.8	+958	+14.4
Public services										
1951–61	+13	+1.0	+54	+7.0	+38	+6.0	+200	+8.0	+302	+6.0
1961–71	+25	+2.0	+170	+21.6	+110	+17.7	+502	+17.3	+807	+14.5
1971–81	−78	−7.4	+102	+8.8	+53	+6.5	+456	+14.1	+488	+7.7
Total employment[a]										
1951–61	+43	+1.0	+231	+6.0	+140	+6.0	+1060	+10.0	+1490	+7.0
1961–71	−643	−14.8	+19	+0.6	+54	+2.4	+1022	+8.5	+320	+1.3
1971–81	−538	−14.6	−236	−7.1	−150	−5.4	+404	+3.5	−590	−2.7

[a] Including primary sector.

Source: Table 2.7, Begg *et al.* (1986).

Table 1.4. *Trends in Residential Unemployment Rates*[a] *Relative to Great Britain, 1951–1981* (Great Britain = 100 in each year)

Type of area	1951	1961	1971	1981
Inner areas of six large conurbations	133	136	144	151
Outer areas of six large conurbations	81	82	88	101
Six large conurbations	103	105	112	122
Seventeen free-standing cities	95	107	112	115
Smaller towns and rural areas	95	93	90	90
Great Britain actual percentage unemployment rates	2.1	2.8	5.2	9.8

[a] Census of Population definition of 'out of work'.

Source: Adapted from Table 2.8 in Begg *et al.* (1986).

recession, with areas of employment growth to be found almost exclusively in the three southern regions—the South East, East Anglia, and the South West (Table 1.6). Restricting the measure of employment growth even further to male employment, only East Anglia and the Rest of the South East (excluding London) experienced employment growth between 1971 and 1981. Recent studies of the performance of Britain's urban areas (Begg and Moore 1986; Champion and Green 1985) have demonstrated that the best-performing areas (as measured by employment growth in one study, and by employment growth, unemployment, population growth, and a surrogate for income in another) were medium-sized cities overwhelmingly located in the south, and particularly in the London metropolitan region, although not including London. The poorest-performing areas were the conurbations and other manufacturing cities and towns of the north, Scotland, and Wales.

Urban Structural Economic Change

Behind urban unemployment and employment growth problems lie some important urban structural economic changes. In addition to regional location, several other structural characteristics are correlated with urban economic performance. Specializations in finance, banking, insurance, and related business services, and in new and more advanced manufacturing industries, are closely related to better economic performance of urban areas. Size of the area (level of urbanization) is not an important factor affecting the performance of areas in the south. In the north the dominance of manufacturing employment and the size of urban areas are significant influences on poorer local economic performance (Champion and Green 1985).

Thus, the economic functions contributing to urban economic success and the location of urban employment growth in Britain have changed along with the changing structure of the national economy. The economic performance

Table 1.5. *Ranking of Most Deprived Areas in Great Britain*

Rank[a]	Area	Population change 1971–81
1	Glasgow old core	−22.0[b]
2	Glasgow peripheral	−22.0[b]
3	Birmingham old core	−19.3
4	Hull core	−35.8
5	Derby core	−20.1
6	Manchester–Salford old core	−25.5
7	Liverpool old core	−23.1
8	Nottingham core	−31.3
9	Teesside core	−15.6
10	Other W. Midlands cores	−8.2
11	Other Strathclyde cores	−22.0[b]
12	Other Greater Manchester cores	−11.5
13	Leicester core	−7.1
14	Merseyside peripheral	−15.3
15	W. Yorkshire cores	−14.4
16	London Docklands	−16.9
17	Plymouth core	−13.1
18	Other Tyne and Wear cores	−17.0
19	Sheffield core	−16.0
20	Newcastle–Gateshead old core	−13.8
21	Other Merseyside cores	−13.2
22	Other S. Yorkshire cores	−3.8
23	Stoke core	−11.5
24	Hull outer area	+6.2
25	Hackney and Islington	−19.1
26	Kensington and Chelsea, Haringey, and Westminster	−15.0
27	Lambeth	−23.0

[a] Lowest ranking is most deprived.
[b] No data on Strathclyde's population change by sub-area could be found on a consistent basis—the value shown is the Glasgow average.

Source: Begg and Eversley (1986).

of an area and its potential for future employment growth are related to its 'specialization' in declining or expanding sectors and its ability to shift its economic base from one to the other (Begg and Moore 1986). Between 1971 and 1981 Great Britain experienced substantial employment decline in traditional manufacturing industries related to the early stages of industrialization, sectors of post-war industrial growth, and capital-intensive sectors, all open to national and international competition. These industries lost approximately 25 per cent of their employment, and by 1981 total employment in goods production constituted 38 per cent of employment in Britain compared to 52 per cent in 1961.[3] Advanced manufacturing lost jobs nationally, but at a slower rate than other manufacturing. In the same period major

Table 1.6. *Regional Employment Change in Great Britain, 1971–1981*

Region	Total (%)	Male (%)	Female (%)
East Anglia	+12.2	+3.4	+25.4
Rest of the South East	+11.2	+2.5	+24.8
South West	+8.2	−1.0	+23.4
East Midlands	+4.0	−3.8	+17.4
Scotland	−0.6	−8.0	+10.9
Great Britain	−1.5	−8.9	+10.5
Wales	−2.6	−13.1	+17.1
Yorkshire and Humberside	−3.9	−11.1	+8.4
North	−7.3	−15.4	+7.4
North West	−7.6	−14.8	+3.7
West Midlands	−7.9	−14.6	+3.5
Greater London	−9.5	−14.2	−2.1

Source: Census of Employment, 1971 and 1981. From Table 4.1 in Buck and Gordon (1986).

employment growth took place in the service sector, with the greatest percentage increase (nearly 40 per cent) in advanced business services. Total employment in service production constituted 62 per cent of British employment in 1981 compared to 48 per cent in 1961.

The cities experiencing the slowest employment growth in the 1970s (Table 1.7) were conurbations and older manufacturing centres in the north which in 1971 had above-average shares of employment in older declining manufacturing industries (particularly mining, metals manufacturing, and shipbuilding) and below-average shares of labour-intensive and more rapidly growing service sectors (particularly business services) and advanced manufacturing industries. Despite the severe contraction of the older manufacturing industries, employment in these northern cities in 1981 was still over-represented in these declining sectors and the cities still had a relatively low proportion of employment in the expanding service sectors. Continued specialization in declining sectors does not bode well for the future growth prospects of these cities. By contrast, the cities experiencing the strongest employment growth in the 1970s (Table 1.8) had in 1971 relatively low proportions of employment in older manufacturing industries and relatively high proportions in advanced business services and post-war growth and 'high-technology' industries. They retained this successful profile in 1981. The best-performing cities are those whose changing economic functions coincide with the expanding sectors of the national economy.

The cities with the highest proportions of employment in the key sectors of advanced manufacturing (for example, pharmaceuticals, scientific instruments, electronic components, and electronic computers) and advanced pro-

Table 1.7. *The Bottom Twenty Declining Cities in Great Britain, 1971–1981* (Total Employment = 100)

City	Index as at 1981		
	Total	Males	Females
Sunderland	68.7	56.1	87.0
Liverpool–Birkenhead	76.3	70.1	86.1
Scunthorpe	77.1	66.2	106.2
Deane Valley	79.0	97.5	79.0
Airdrie–Coatbridge	79.4	66.7	101.7
Huddersfield	80.8	77.9	85.8
Kettering	82.2	69.9	103.2
Thames Estuary	82.6	74.6	100.4
Coventry–Nuneaton	82.7	73.8	100.5
Glasgow	83.6	76.6	94.1
Newport (Wales)	84.5	73.9	107.6
Tyneside	84.9	78.5	95.3
Blackburn	85.2	80.2	93.1
Birmingham	86.1	81.0	95.0
Hartlepool	86.3	79.4	98.0
Sheffield	86.4	78.9	99.6
Greenock	86.7	78.4	101.5
The Potteries	87.9	84.1	93.7
Cardiff	88.6	79.0	104.1
Motherwell	88.6	84.5	98.7
Average of bottom 20 cities	82.9	76.4	96.5
Great Britain	97.3	90.0	109.0
Average of bottom 20 cities relative to GB	85.2	84.9	88.5

Source: Table 3.7, Begg and Moore (1986).

ducer services reveal the changing urban and regional location of economic growth (Tables 1.9 and 1.10). These cities contrast with the older industrial towns of the north which dominate the rankings in the traditional manufacturing sectors. The advanced manufacturing sector in 1981 had a significant representation of cities in the south of the country and New Towns in its top twenty. Similarly the top twenty cities in producer services in 1981 include only four northern cities, while the poorest-performing cities in this sector were uniformly declining industrial areas.

Growth in producer services has become a critical independent factor in urban employment growth and has altered the hierarchy of urban economic performers in Britain. Older manufacturing centres have not developed significant concentrations of producer services, and some may even have experienced the contraction of existing services as a result of the decline in manufacturing and the expansion of branch-plant economies (Hall 1985). These findings echo those of Noyelle and Stanback (1984) on the significant

Table 1.8. *The Twenty Fastest-growing Cities in Great Britain, 1971–1981* (Total Employment = 100)

City	Index as at 1981		
	Total	Males	Females
Milton Keynes	189.7	177.9	210.1
Aberdeen	137.4	139.2	134.7
Basingstoke	136.3	121.0	162.3
Cheltenham	133.8	127.8	142.8
Basildon	127.7	120.4	139.6
Reading	124.8	120.1	131.5
Aldershot	121.3	108.9	140.0
Northampton	119.5	110.6	132.3
Peterborough	118.8	107.6	142.3
Eastbourne	118.4	112.0	125.6
Maidstone	117.1	110.2	128.3
High Wycombe	116.0	104.8	134.0
Cambridge	115.5	109.1	124.9
Bedford	114.4	102.0	134.9
Exeter	114.2	102.0	134.9
Harrogate	113.6	105.6	124.1
Hertford	113.5	105.6	124.1
Colchester	112.6	103.5	127.1
Southend	111.0	102.5	120.2
Crawley	110.7	102.7	123.3
Average of top 20 cities	123.3	114.5	137.2
Great Britain	97.3	90.0	109.0
Average growth of top 20 relative to GB growth	126.7	127.2	125.8

Source: Table 3.4, Begg and Moore (1986).

influence of service sector growth on the urban system of the United States of America (USA). Among the fastest-growing US metropolitan areas in the 1960s and 1970s were newer 'Sunbelt' cities and restructured older large northern cities with economic roles as diversified or specialized service centres. Cities with corporate management, finance, and professional services replaced production centres as the top economic performers. Resort areas, centres of government and non-profit services, and military–industrial complexes were also among the fastest-growth areas. Pure production centres and production centres with corporate headquarters were the worst-performing metropolitan areas. The rise of exporting service-based urban economies is linked to changes in corporate management, namely the greater importance of 'control' and 'development' functions which require an increasing share of expenditure on service inputs to product development and marketing, research and development (R&D), and corporate strategy and planning. To service this 'administrative economy', educational, cultural, recreational,

Table 1.9. *Employment in the Top and Bottom Twenty Cities in Advanced Manufacturing as a Percentage of Total Employment in 1971 and 1981*

Top twenty						Bottom twenty					
Rank		% 1971		% 1981	Rank 1971	Rank		% 1971		% 1981	Rank 1971
1	Chelmsford	23.9	Hertford	19.5	2	20	Falkirk	1.1	Peterborough	1.2	—
2	Hertford	22.1	St Albans	17.2	3	19	Sheffield	1.0	Cardiff	1.2	—
3	St Albans	21.6	Derby	15.4	6	18	Harrogate	1.0	Morecambe	1.2	8
4	Hamilton and E. Kilbride	18.2	Chelmsford	13.2	1	17	Maidstone	0.9	Teesside	1.2	—
5	Harlow and Bishop's Stortford	17.7	Coventry and Nuneaton	12.8	7	16	Mansfield	0.8	Aberdeen	1.2	12
6	Derby	17.4	Greenock	11.0	—	15	Chester-le-Street	0.8	Maidstone	1.1	17
7	Coventry and Nuneaton	17.3	Hemel Hempstead	10.6	17	14	Telford	0.7	Chester	1.0	—
8	Basildon	16.5	Portsmouth	10.3	—	13	Swansea	0.5	Swansea	0.9	13
9	Airdrie and Coatbridge	14.7	Hamilton and E. Kilbride	10.3	4	12	Aberdeen	0.5	Barnsley	0.9	—
10	Guildford	13.4	Preston	9.7	—	11	Kettering	0.4	Warrington	0.8	6
11	Cheltenham	11.9	Chatham	9.6	—	10	Tamworth	0.4	Scunthorpe	0.8	1
12	Hartlepool	11.3	Crawley	9.6	14	9	Burton upon Trent	0.4	Wakefield and Dewsbury	0.7	—
13	Basingstoke	11.2	Basingstoke	9.5	13	8	Morecambe	0.4	Falkirk	0.7	20
14	Crawley	11.2	Harlow and Bishop's Stortford	9.5	—	7	Newport	0.3	Mansfield	0.5	16
15	Bristol	9.3	Guildford	9.4	10	6	Warrington	0.3	Doncaster	0.5	4
16	Brighton	9.0	Gloucester	9.3	—	5	Darlington	0.3	Blackpool	0.4	—
17	Hemel Hempstead	9.0	Reading	8.9	—	4	Doncaster	0.2	Carlisle	0.3	—
18	Swindon	8.9	Cheltenham	8.0	11	3	Cannock	0.0	Grimsby	0.2	—
19	SE Fife	8.9	Basildon	8.8	8	2	Dearne Valley	0.0	Darlington	0.2	5
20	Aldershot	8.8	Burnley	8.6	—	1	Scunthorpe	0.0	Dearne Valley	0.1	2
Great Britain %	4.9 (1971)	4.4 (1981)									

Source: Table 3.A5, Begg and Moore (1986).

Table 1.10. *Employment in the Top and Bottom Twenty Cities in Producer Services as a Percentage of Total Employment in 1971 and 1981*

Top twenty						Bottom twenty					
Rank		% 1971		% 1981	Rank 1971	Rank		% 1971		% 1981	Rank 1971
1	Aldershot	19.1	Crawley	19.6	2	20	The Potteries	2.3	Grimsby	4.1	—
2	Crawley	13.9	Aldershot	18.8	1	19	Doncaster	2.3	Hamilton and E. Kilbride	4.1	—
3	London	13.7	London	17.3	3	18	Sunderland	2.3	The Potteries	3.9	20
4	Harrogate	9.7	Reading	13.6	7	17	Chesterfield	2.2	Scunthorpe	3.9	13
5	Bedford	9.5	Chester	12.3	—	16	Swindon	2.1	Derby	3.8	—
6	Edinburgh	9.2	Southend	12.1	17	15	Wakefield and Dewsbury	2.1	Dearne Valley	3.8	1
7	Reading	8.8	Edinburgh	12.0	6	14	Tamworth	2.0	Blackburn	3.5	—
8	Cheltenham	8.3	Brighton	11.8	11	13	Scunthorpe	2.0	Chesterfield	3.4	17
9	Blackpool	8.2	Norwich	11.7	10	12	Greenock	2.0	Morecambe	3.3	—
10	Norwich	7.9	Cheltenham	11.5	8	11	Thames Estuary	1.9	SE Fife	3.2	—
11	Brighton	7.7	Guildford	11.4	14	10	Chester-le-Street	1.9	Wakefield and Dewsbury	3.1	15
12	Harlow and Bishop's Stortford	7.6	Slough	10.5	—	9	Hartlepool	1.8	Falkirk	3.1	8
13	Cambridge	7.6	Bristol	10.3	—	8	Falkirk	1.7	Burton upon Trent	3.0	—
14	Guildford	7.5	Warrington	10.2	16	7	St Helens	1.6	Doncaster	3.0	19
15	Bournemouth	7.5	High Wycombe	10.1	—	6	Barnsley	1.6	Greenock	3.0	12
16	Warrington	7.3	Bournemouth	10.1	15	5	Motherwell and Wishaw	1.5	Motherwell and Wishaw	2.9	5
17	Southend	7.1	Ipswich	10.1	—	4	Airdrie and Coat-bridge	1.3	Mansfield	2.8	3
18	Northampton	6.5	Northampton	9.8	18	3	Mansfield	1.2	Hartlepool	2.5	9
19	Exeter	6.1	Luton	9.7	—	2	Cannock	1.1	Barnsley	2.4	6
20	Hertford	6.0	Blackpool	9.7	9	1	Dearne Valley	0.9	Cannock	1.9	2
Great Britain % 6.1 (1971) 8.5 (1981)											

Source: Table 3.A6, Begg and Moore (1986).

and other services grew (Smith and Williams 1986). The US research expresses concern that a dichotomy is growing between service centres with control over major investment and employment decisions and cities dependent on those decisions which will leave the latter group with little control over their own economic development. Policy-makers must be concerned with whether the factors contributing to the local expansion of producer services are amenable to policy intervention.

However, we must not discount the continued importance of the manufacturing sector to local economic growth. Wolman's research (Wolman 1986) corroborates the findings of Begg and Moore that those cities whose industrial structures were dominated by manufacturing employment at the beginning of the 1970s experienced poor economic performance during the decade of the 1970s. In Britain, however, the difference between successful and unsuccessful urban economies was related at least as much to differences in growth in the manufacturing sector as to differences in growth in the service sector, and the performance of the manufacturing sector had a very strong influence on changes in urban employment and unemployment. The manufacturing sector can make an important contribution to urban economic success. However, in Britain, the manufacturing sector performed less well where manufacturing had dominated the urban economy—that is, in older industrial centres. (Interestingly, in his international comparative study, Wolman found that the dominance of manufacturing in German cities did not on the whole have a negative effect on urban economic performance, as it did in the USA and Britain.) Thus, modern advanced manufacturing industries have disproportionately grown in southern England. Wolman suggests that the service sector did not have as great an influence on urban economic performance in Britain as it did in the USA, because the British service sector has not performed as well as that in the USA, growing at approximately half the rate of the US service sector between 1971 and 1981.

Behind the geographic shifts in economic activity and performance lies the fact that 'locational preferences for a wide range of economic activity have changed profoundly' (Hall 1981, 43). The comparative advantages of urban centres are changing for the different types of economic activity which originally contributed to the growth of the nation's large urban centres. These centres are losing their attraction for capital-intensive manufacturing, routine office functions, warehousing, and distribution; public sector services are declining; and personal consumer services are decentralizing along with the out-migration of population. The decentralization of routine office functions, facilitated by advances in information technology, has been the major contributor to the growth of the M4 Corridor rather than the growth of high-technology firms (Breheny *et al.* 1983). Urban centres remain attractive to: higher corporate management functions; advanced services to business, including finance, legal, accounting, advertising, and consultancy; and education and government services. However, the older urban centres face

Table 1.11. *Changes in Total Employment in Great Britain, including the Five City Study Areas, 1951–1981*

Type of Area	1951	1961	1971	1981
Inner areas of six large conurbations	100	101	86	74
Outer areas of six large conurbations	100	107	109	99
Six large conurbations	100	104	99	88
Greater London	100	105	95	84
West Midlands	100	109	107	93
Tyneside	100	105	103	94
Clydeside	100	100	91	78
Seventeen free-standing cities	100	106	105	100
Bristol	100	105	108	106
Smaller towns and rural areas	100	111	116	118
Great Britain	100	108	107	103

Note: The figures show total employment (index, 1951 = 100).

Source: Adapted from Table 2.9 in Begg *et al.* (1986).

important disadvantages with regard to the expanding producer services sector, as these services are more mobile than traditional manufacturing industries. Since these sectors employ larger numbers of professional staff, their locational decisions 'are likely to be based far more on satisfying the residential preferences of their staff than on more conventional characteristics like transport costs, or proximity to markets' (Begg and Moore 1986, 49).

Comparative Economic Performance of Five Urban Areas

The trends of urban economic change and their related problems of rising unemployment and increasing concentrations of unemployment and deprivation are reflected in the experiences of all the city study areas. However, there is significant diversity in their economic performance relative to each other, other cities in their size groups, and the nation. The six large conurbations as a group experienced a long-term decline of 12 per cent in total employment between 1951 and 1981, while the free-standing cities only maintained their employment; however, in the same period, the nation increased its employment, and smaller towns and rural areas improved their share of employment (Table 1.11). Relative to the nation, only Bristol among the five city study areas had a moderately successful performance between 1971 and 1981. Great Britain's employment fell by 3.6 per cent while Bristol's dropped by only 3.2 per cent (Table 1.12). Using Functional Urban Regions (FURs), which encompass entire urban labour markets, Bristol's employment change was even better between 1971 and 1981 (Table 1.13): total employment declined by only 0.74 per cent, a performance that was nearly 2 percentage points better than that of the nation (Wolman 1985). Bristol also outperformed the free-standing cities as a group, although it was

Table 1.12. *Percentage Employment Changes in the Five City Study Areas, 1971–1981 and 1951–1981*

Area	1971–81	1951–81
West Midlands	−13.9	−9.0
Clydeside	−14.7	−24.8
Bristol	−3.2	5.5
Greater London	−11.7	−11.8
Tyneside	−8.8	−14.2
Great Britain	−3.6	3.4

Source: Adapted from Begg *et al.* (1986), Tables 3.A1–3.A9.

Table 1.13. *Total Employment Change in the Five City Study Areas, 1971–1981, based on FURs*

Area	% change 1971–81	1981 index (1971 = 100)	1981 index relative to all FURs	1981 index relative to GB
West Midlands	−11.73	88.27	0.88	0.91
Bristol	−0.74	99.26	0.99	1.02
Clydeside	−13.20	86.80	0.87	0.89
London	−8.87	91.13	0.91	0.93
Tyneside	−8.31	91.69	0.92	0.94
61 FURS[a]	−0.23	99.77		1.02
Great Britain	−2.52	97.48		

[a] All urban areas with 1971 populations in excess of 200 000.

Source: Wolman (1985).

not one of the only four cities in this group which did not lose employment during the decade. Over the longer term, 1951–1981, Bristol's stronger economic performance was more evident. During the thirty years, Bristol increased its employment by 5.5 per cent, compared to a 3.4 per cent increase for the nation.

By contrast, total employment in the other four city study areas fell substantially, with Clydeside's performance consistently the worst but with the West Midlands conurbation approaching a similar level of decline in the 1970s. The West Midlands conurbation and Bristol experienced similar employment growth between 1951 and 1971, before the performance of the Birmingham area plummeted in the 1970s. With regard to employment of its residents, Bristol outperformed the free-standing cities as a group until 1961, but then deteriorated faster than its group, ending the three decades very much on a par with other cities of its size (Table 1.14).

The employment performance of the five city study areas also demonstrates the economic difficulties of the nation's largest urban areas. While

Table 1.14. *Changes in Residential Employment in Great Britain, including the Five City Study Areas, 1951–1981*

Type of area	1951	1961	1971	1981
Inner areas of six large conurbations	100	96	74	57
Outer areas of six large conurbations	100	105	104	96
Six large conurbations	100	101	92	79
Greater London	100	104	93	78
West Midlands	100	107	100	85
Tyneside	100	99	93	84
Clydeside	100	99	85	73
Seventeen free-standing cities	100	100	97	88
Bristol	100	104	100	88
Smaller towns and rural areas	100	118	120	124
Great Britain	100	108	107	103

Note: The figures give residential employment (index, 1951 = 100).
Source: Adapted from Table 2.13 in Begg *et al.* (1986).

only the performance of Bristol among the city study areas was better than the nation's between 1971 and 1981, the sixty-one large urban areas (FURs) of the nation (that is, those with a population in 1971 greater than 200 000) as a group outperformed the nation and experienced an even more negligible decline in employment than the Bristol urban region (Wolman 1985) (see Table 1.13).

With regard to unemployment among urban residents, the six large conurbations as a group and the large free-standing cities have both experienced a gradually deteriorating situation relative to the nation and small towns and rural areas. However, the experiences of the individual city study areas have differed (Table 1.15). Tyneside and Clydeside had very high unemployment throughout the three decades 1951–81. The residential unemployment rate of the West Midlands conurbation steadily declined from a very good position relative to the nation in 1951, and then collapsed in the 1970s from just below the national average to 38 per cent above it in 1981. Unemployment in Bristol, which had been 24 per cent below the national average in 1951, worsened substantially in the 1970s, nearly reaching the national average by 1981. Only Greater London maintained its long-term favourable unemployment situation relative to the nation. However, if unemployment rates are examined for travel-to-work areas (FURs), then both Greater London and the Bristol urban region had unemployment rate increases from 1971 to 1981 well below the increase in the national rate (Wolman 1985) (Table 1.16).

Thus, among the five city study areas, only the Bristol urban region's economic performance, as measured by employment growth and unemployment, was a successful one during both the 1971–81 and the long-term 1951–81 periods. However, despite the success, Bristol's unemployment

Table 1.15. *Changes in Residential Unemployment Rates[a] Relative to Great Britain, 1951–1981, including Data for the Five City Study Areas* (Great Britain = 100 in each year)

Type of area	1951	1961	1971	1981
Inner areas of six large conurbations	133	136	144	151
Outer areas of six large conurbations	81	82	88	101
Six large conurbations	103	105	112	122
Greater London	90	82	88	89
West Midlands	57	79	98	138
Tyneside	195	157	148	140
Clydeside	195	193	158	173
Seventeen free-standing cities	95	107	112	115
Bristol	76	89	77	97
Smaller towns and rural areas	95	93	90	90
Great Britain actual percentage unemployment rates	2.1	2.8	5.2	9.8

[a] Census of Population definition of 'out of work'.

Source: Adapted from Table 2.8 in Begg *et al.* (1986).

Table 1.16. *Percentage Unemployment Rates for 1971 and 1981 for the Five City Study Areas, based on Travel-to-work Areas*

Area	1971	1981	% point change	Change relative to all FURs	Change relative to GB
West Midlands	3.8	13.3	9.5	1.20	1.27
Bristol	2.9	9.0	6.1	0.77	0.81
Clydeside	7.4	15.2	7.8	0.99	1.04
London	1.8	7.1	5.3	0.67	0.71
Tyneside	5.4	13.5	8.1	1.03	1.08
61 FURs	3.7	11.6	7.9		1.05
Great Britain	3.4	10.9	7.5		

Source: Wolman (1985).

deteriorated during the 1970s relative to the nation, other large free-standing cities, and small towns and rural areas. However, along with London, Bristol has weathered the recent recession better than the nation, with a smaller rise in unemployment. In addition, Bristol as a whole does not share the inner city problems of inner London and the other conurbations, despite its significant concentrations of deprivation; nor does it share the extensive dereliction resulting from the decline of traditional heavy manufacturing. Nevertheless, the Bristol 'success' is a mixed one.

The other southern study area, Greater London, has also been relatively successful. Despite its substantial long-term decline of population, a 16 per cent fall in employment between 1951 and 1981 (encompassing both manufacturing and service sector employment), employment decline more severe than national trends and that of other cities, and the spread of manufacturing employment decline throughout the conurbation, London's unemployment rate remains below that of the nation, and its individual earnings remain high. Its manufacturing employment performance during the recession has been better than that of any other conurbation. However, London's inner city unemployment is very high and has worsened significantly during the recession. London's problem is an inner city problem of concentrated deprivation and unemployment.

In contrast to London and Bristol, Glasgow, Tyneside, and the West Midlands urban area were among the twenty cities in Britain that experienced the most rapid employment decline between 1971 and 1981 (Begg and Moore 1986). That group of cities (as defined by Begg and Moore) with the poorest economic performance also includes several others in the urban regions studied by the ESRC Programme: Motherwell, Sunderland, and Coventry. The northern industrial areas of Tyneside, Clydeside, and the West Midlands conurbation all face metropolitan-wide and not just inner city economic problems. Between 1951 and 1981 they lost employment while the nation gained, and in the 1970s they lost employment faster than the nation and their unemployment rates increased faster than the nation's. Clydeside experienced the longest and worst decline while the performance of the West Midlands conurbation soured rapidly in the 1970s. The national earnings ranking of the West Midlands conurbation dropped precipitously, and it now has one of the highest long-term unemployment rates in the nation.

The city study areas have also differed with regard to the structure of their economies and success in the adjustment of those structures to economic change. Successful urban economic performance is closely correlated with the proportion of expanding economic sectors in the local economy and the ability of an area to adjust from declining to expanding sectors (Begg and Moore 1986). This is evident in the anatomy of Bristol's relative economic success. Despite a significant proportion of its employment being in manufacturing in 1971 (comparable in percentage terms to the proportions in Glasgow and Tyneside) and a substantial decline in manufacturing employment in the 1970s (more rapid than the national manufacturing decline between 1971 and 1978), the Bristol urban region experienced a smaller percentage decline in manufacturing employment as well as a greater increase in service sector employment than the other four study areas and the nation between 1971 and 1981 (Wolman 1985). Moreover, uniquely among the study areas, substantial decline in Bristol's manufacturing employment was numerically replaced by substantial increases in service sector employment, particularly in insurance and business and consumer services. Most importantly, by 1981 Bristol had become one of the twenty cities in Britain with the highest proportion of employment in the most rapidly growing service sector, producer services (see Table 1.10).

Bristol's relatively successful structural adaptation indicates the replacement of older manufacturing cities at the top of the economic performance league by cities with roles in housing corporate headquarters and higher-level corporate functions such as R&D; in the export of business services such as Bristol's rapidly expanding financial services sector; and in technologically advanced industries such as Bristol's aerospace industry. Despite its high-technology image, Bristol has not yet mimicked the Silicon Valley experience of rapidly expanding computer and electronics companies and 'mushrooming new enterprises' or become a location for high-technology production as Scotland has. Bristol has also had only limited growth in non-aerospace high-technology employment. As already noted, office development has been a far greater contributor to the M4 Corridor's growth than high-technology industry. The Bristol experience also bears out the findings of other research on the importance of defence and defence-related R&D and production in the growth of the high-technology sector in area economies in the USA (Saxenian 1983; Markusen 1984, 1985).

London has also been most successful in attracting employment in the advanced business services sector. The high share of employment in this sector, and from its role as a national and regional centre of public and other services, distinguishes London from other conurbations. Occupationally, Bristol and London have a significantly smaller proportion of semi-skilled and unskilled employment and a higher proportion of professional and other skilled employment compared to the three older industrial areas.

By contrast to Bristol and London, the rapid decline of employment in Clydeside, Tyneside, and the West Midlands during the 1970s is closely related to the domination of their economies by rapidly declining traditional manufacturing industries. In 1971 the Clydeside and, particularly, the Tyneside economies 'specialized' in that segment of the older industries sector dependent on particular relatively immobile resources found only in a limited number of locations: for example, metal manufacturing, coal-mining, and shipbuilding. Despite the decline of employment in these industries, by 1981 Clydeside and Tyneside had increased their concentration in this sector, while both advanced manufacturing and the business services sector were seriously under-represented. In 1971, of the economies of all three conurbations, that of the West Midlands was the most heavily dominated by both the older and newer classes of traditional manufacturing industry, while also having low shares of employment in modern manufacturing and producer services. This sitution was unrelieved in 1981. The older industrial conurbations of the West Midlands and the North have not yet been able to establish new, diversified, and modernized economies attuned to growth sectors in services or manufacturing.

Explanations for Change

The city studies offer a number of explanations for the failure of old industrial cities to maintain employment levels and adapt to change. They place particular emphasis on the domination of an area's economy by traditional manufacturing industry and the related local structural characteristics, costs of production, and environmental constraints which impair the competitiveness of existing firms and impede the growth of new economic activity. The influence of these factors is enhanced by the declining competitive advantages of urban locations and the non-urban residential preferences of professionals and technicians in growth sectors.

First, there is the importance of the inherited industrial structure and the processes of corporate restructuring to address declining business competitiveness. Tyneside and Clydeside, dominated by traditional heavy manufacturing industry (shipbuilding, coal-mining, and iron and steel), or the West Midlands, with a more recent generation of heavy industry (for example, mechanical engineering), have performed poorly. Their industries have been severely affected by increasing international competition, historical patterns of low investment, and declining productivity and competitiveness. The resulting process of corporate restructuring which addressed these changes in the competitive environment involved closures, rationalization, and the introduction of labour-saving new technology that reduced employment in these communities, while new investment aimed at the diversification and internationalization of production has been attracted to less urbanized and industrialized areas in the UK and abroad. The restructuring strategies of large manufacturing firms have negatively affected the economies of the old industrial cities and regions, as these firms have made decisions to locate away from those areas where firms obtain lower rates of return on capital. The continued competitive pressures to increase productivity and reduce costs in the now smaller manufacturing base do not augur well for an employment revival in this sector. The health of the sector, however, will affect the level of employment that can be retained in manufacturing as well as the growth of employment in the service sector.

The dominance of manufacturing in the urban economy is a major contributory factor to the recent problems of Britain's poorly performing cities (Wolman 1986). However, the city studies indicate that the poor economic performance of the Clydeside, Tyneside, and West Midlands conurbations, and of their inner areas, cannot be adequately explained by the fact that their employment was disproportionately concentrated in nationally declining industrial sectors. Inner city areas performed far worse, and outer cities and towns better, than predictions based on their economic structures. Local economic factors—that is, characteristics of local firms and industries and of the local area which handicap business performance—are important to the explanation of urban economic decline, although the industrial structure of these urban areas has had a significant negative influence on these factors.

The three old industrial city study areas manifest important patterns by local firms of low investment, poor product innovation, older, more labour-intensive capacity, and poor premises, high labour costs, inefficient work practices, and poor industrial relations, routine production of mature products, and low-value-added activity, all resulting in the declining productivity, competitiveness, and profitability of firms. Thus firms in these areas were more vulnerable to rationalization and closure. In the West Midlands conurbation, the dependence of the area's economy on a small group of large firms in a few older industries with close linkages to many local firms exacerbated the consequences of decline in these industries. In Tyneside, the growth of the external control of firms and establishments contributed to production activities which were of marginal importance to firms headquartered elsewhere and therefore vulnerable to contraction and closure, and to limitation of the growth of business services which assist corporate management functions. However, the distinction between foreign and domestic external control does not appear to be significant.

The domination of an area's economy by heavy manufacturing industry also has detrimental effects on the generation of new economic activity. As one analyst has argued recently, the structure of older industrial areas at the end of an economic 'life cycle' is marked by inflexibilities in supply-side characteristics (for example, skills, mobility, entrepreneurship, and innovation) which create unfavourable conditions for new economic growth (Steiner 1985). Other areas with fewer barriers which can provide the requisite supply-side factors at lower cost are attracting new economic growth. The inner cities of the northern study areas suffer from low rates of new firm formation. In addition, the Newcastle study suggests that the types of new businesses established in the area are less likely to grow as a result of the effects of the area's industrial structure, low levels of wealth, and high proportion of blue-collar workers on entrepreneurial attitudes.

However, the studies raise questions about the importance of small firms to local economic growth. In all five city study areas, but particularly Bristol, the smallest size sector of firms for which data were available (11–50 employees) had higher rates of employment growth than any other size sectors. However, the absolute contribution of the sector to employment growth was very modest given its low base level, and therefore did not compensate for the loss of manufacturing employment. Bristol's relative success is not the product of a small-firms-oriented economy. In the mid-1970s Bristol had an employment structure more dominated by large firms (500+ employees) and a smaller percentage of employment in small firms than all the other study areas. Significantly, Bristol's large manufacturing firms, while experiencing significant employment losses (their contraction accounted for the bulk of manufacturing employment losses), performed better than the large-firms sector of the other study areas. At the same time, the dominance of large firms has not prevented small firms from making a positive, if only modest, contribution to economic growth in the Bristol region, in contrast to the old industrial cities.

The problems of declining manufacturing and certain service industries located in inner cities were exacerbated by problems with local factors of production, including the higher costs of, or the unavailability of, land, premises, and skilled labour. Higher operating costs affect the profitability and location decisions of inner city firms. Labour costs have become a more important factor of production. The West Midlands study maintains that higher costs, weakness in training, a dearth of 'new skills' which would facilitate the adoption of new technologies, relatively poor productivity, and militancy against changes in the labour process have made labour in old manufacturing cities uncompetitive relative to that in other areas. The lack of land for expansion was also a contributory factor in industrial decline in the West Midlands. Similarly, London's loss of inner city manufacturing employment has in part resulted from increasingly capital-intensive manufacturing being crowded out of the urban land market by competition from the office and residential sectors. Higher costs and physical constraints have also contributed to the out-migration of routine and even higher office functions—benefiting areas such as Bristol. In Clydeside, land constraints were not a factor, although there is strong demand for small units. The agglomeration economies which supported the earlier rise of urban areas have now been replaced by the disincentives of age, obsolescence, disrepair, and congestion supported by changes in communications and transportation which facilitate the decentralization of economic activity. Decline which results in lower-cost city sites may offer profitable development opportunities. However, whether these opportunities are grasped will depend on alternative investment opportunities, the investment climate, and public sector support for urban development.

Policy-makers must be concerned with strengthening the variety of local factors which provide the foundation for long-term employment growth and economic development and facilitate the process of economic adjustment.

The city studies indicate that manufacturing decline is not the only major contributor to urban employment loss. The decline of employment in the distribution and consumer services sectors, and in public services affected by reductions in public expenditure, has also had significant effects. For example, in Glasgow only a small percentage of the unemployed came from metals, mechanical engineering, and shipbuilding, while the entire conurbation had significant losses of employment from distribution, personal services, public administration, and even professional and scientific services. Increases of employment in banking and insurance and business services have not compensated for these employment losses, particularly the loss of full-time employment for male manufacturing workers and semi-skilled and unskilled manual labour. The changing competitive advantages of inner city locations for certain economic activities, the effects of new technology, the out-migration of business and population, and constraints on public spending have had important negative effects on urban employment.

But what of the factors contributing to relative local economic success? In the contemporary context of corporate restructuring to adjust to change, Bristol has offered a number of locational advantages to a range of economic functions and has thereby fostered the retention, expansion, and location of firms and employment growth in the area. Substantial public expenditure on defence procurement in Bristol's aerospace industry has lessened the decline of the manufacturing sector and helped to establish a core of advanced-technology activity. The aerospace industry has also promoted other high-technology and specialist services through its purchasing and subcontracting, and strengthened the area's pool of skilled labour and technical expertise. However, defence expenditure has not produced the significant regional agglomeration economies which have occurred in defence-dependent regions of the USA, and therefore has not become a motor for regional transformation (Boddy *et al.* 1986).

Other factors have also contributed to Bristol's success in consolidating its role in higher corporate technical and management functions, the office sector, financial and business services, and the growth of personal consumer services. Most importantly, Bristol has benefited from being a large independent labour market with lower-cost amenable workers and a diversity of skills relevant to administrative, technical, and production functions. In addition, its role as a regional financial and business centre, transportation access to London while being outside the higher costs of its property and labour markets (thus providing attractions to routine office functions), the availability and lower cost of land and premises, and the quality of its environment and its image providing 'relocation appeal' have been important factors. Significant in the Bristol experience is its demonstration of the importance of 'quality-of-life' or 'amenity' factors such as housing, environment, leisure facilities, culture, and personal services to the recruitment and retention of technical, scientific, managerial, and professional staff, and therefore their importance to the location of contemporary growth sectors of advanced manufacturing, corporate control functions, and business services.

London has benefited from an industrial structure much more favourable than those of the other conurbations—although this has not protected it from substantial manufacturing and, more recently, service sector employment decline resulting from cost, physical, and labour constraints and the out-migration of population. Historically, its economy has been characterized by small-scale consumption and service-related manufacturing. Its role as the nation's centre of corporate management, government, finance, business and consumer services, culture, and tourism has provided the strength in the service sector. However, in the face of its several constraints, London is likely to witness continued decline in manufacturing, population-related services, and routine office functions.

Labour Market Adjustment

The Inner Cities Research Programme has highlighted the differential effects of employment change on areas within an urban region and groups within the labour force, and particularly the plight of inner city areas, peripheral estates, and disadvantaged urban residents. Labour market processes have had a particularly strong effect on inner city unemployment. The combination of substantial, but socially selective, out-migration of population from inner cities, increases in the size of the active urban labour force, and a doubling of inward commuting between 1951 and 1981 have concentrated people disadvantaged in the competition for employment in the inner cities, and diverted the reduced number of inner city jobs to outer area residents. Success in maintaining city employment in the face of economic change is unlikely to prevent rapidly rising unemployment, particularly in inner city areas. For example, despite its economic successes, Bristol has significant inequalities in employment opportunities (though more modest in aggregate than the other city study areas): youth, ethnic minority, and long-term unemployment reaching national levels; pockets of very high unemployment in the inner city and peripheral estates; and the replacement of male skilled full-time manufacturing employment by more poorly remunerated female service employment, much of it part time.

More significant than employment decline in explaining disproportionately high rates of inner city unemployment is the fact that certain inner city residents are disadvantaged as a result of personal characteristics, which affect assessments of their employability (Buck *et al.* 1986). These include the young, the unskilled, ethnic minorities, those without formal qualifications, and those with lower educational achievements. In a period of labour surplus, corporate recruiters have used such factors as the individual's employment 'track record', location of residence, and other factors not related to technical competence to filter out applicants for employment. Similarly, public and private employment agencies have eliminated from consideration people with certain personal characteristics. These disadvantaged groups have been disproportionately affected by recession and rising unemployment (particularly long-term unemployment), thus increasing existing inner city disparities. The unemployed are geographically concentrated in inner city areas and peripheral estates because of their residential segregation in council rented housing; and little effort has been made to encourage geographic movement within the council rented sector. Population out-migration from inner city areas, mainly motivated by the search for better housing, has been socially selective—that is, restricted to those who can afford to improve their circumstances. Thus, urban policy is confronted with the problems of spatially concentrated unemployment and multiple deprivation.

However, those disadvantaged in the labour market do not benefit significantly from local employment growth in a situation of depressed employment growth in surrounding areas, as, for example, in northern New Towns (Buck and Gordon 1986). The urban labour market distributes employment opportunities, including those in the inner cities, to those who are more competitive in the labour market. Thus, commuters and in-migrants from surrounding areas capture much new inner city employment. Because of this weak link between local employment growth and unemployment, the targeting of employment creation policies on inner city areas does not necessarily result in a better distribution of employment opportunities to disadvantaged inner city residents. The disadvantaged are only significant beneficiaries of local employment growth within a regional context of employment growth, where they experience lower rates of unemployment, shorter durations of unemployment, improved earnings levels, and greater occupational mobility out of unskilled jobs. Public policies to assist the disadvantaged, such as training, are also more successful in areas of employment growth. Thus, regional policies which could effectively foster employment growth would be an important adjunct to urban economic policies. A policy of encouraging the mobility of disadvantaged inner city residents is also only sensible in a context of regional employment growth—that is, in southern England. In any case, the employment problems of disadvantaged urban residents must be addressed directly.

The Bristol and London studies also indicate that structural economic change is threatening to lead to increasing polarization in the nation's labour markets, between a core group of qualified employees with stable better-paying employment with advancement opportunities, and a larger group in less secure employment with poorer conditions and opportunities in the growing private service sector and in manual labour. The decline of larger manufacturing firms, the demand for qualifications in white-collar producer services, and the constraints on local public expenditure have reduced the availability of more stable public employment and construction work for the less-qualified. They are increasingly dependent for employment on the small, private personal services sector which is characterized by instability of employment and high labour turnover.

The Effects and Effectiveness of Public Policies

The ESRC Inner Cities research suggests that public policies have had important effects on urban economic change. First, and very importantly, there are the unintended effects of public policies other than area development policies. The Bristol area—a 'policy off' area largely excluded from central government regional and urban policy—has nevertheless been a beneficiary of substantial defence and civil expenditure related to its aerospace industry, as well as transportation policies. Defence and civil aviation policies, functioning as an 'implicit' regional policy, have strengthened the area's manufacturing and technical base and contributed to the development of its skilled labour force. These policies have more than compensated the area for the loss of government development assistance.

These findings suggest that central government policies, which have demonstrated their significance to area economic development in growth sectors, should be mobilized to contribute to urban and regional economic development if public policy is to be equal to the scale of efforts required for the economic regeneration of distressed urban regions. For example, defence-related R&D and procurement, as well as other government support for R&D in universities and firms, could make important contributions to strategies to foster high-technology industrial growth in distressed older industrial regions. The Bristol study suggests that such a strategy might be founded on promoting commercial development linked to an area's existing defence specialists.

In contrast, central government policies with regard to local finance, housing, and planning have contributed to increasingly concentrated deprivation in the cities, reducing employment opportunities for disadvantaged urban residents, and restricted urban development. Central government reductions in general financial assistance through rate-support grants to distressed urban authorities and constraints on local capital expenditures have restricted funds for urban development, shifted the tax burden to local residents and businesses, reduced public sector employment for less advantaged urban residents, and depressed the level of local economic activity (Kirwan 1986). Between 1980–1 and 1982–3, rate-support grant assistance to the Urban Programme's highest-priority areas, the Partnership authorities, fell by 10.3 per cent. By 1984, Birmingham City in net terms was £40 million a year worse off as a result of cuts in rate-support grant than when the Birmingham Partnership began in 1978 (Buck *et al.* 1986). Central government restrictions on local finance reduce the effectiveness of local fiscal policy as an instrument of urban economic development. Furthermore, the local rates system does not provide an adequate incentive to local authorities to pursue development policies in order to increase their revenues as a result of expanding the local economic base.

Despite the influence of housing policy on the inner city environment, the social composition of urban areas, the mobility of inner city residents, and urban employment creation, it has not been integrated with urban policy (Ermisch and Maclennan 1986). In the past, the combination of the private housing market and council housing policy contibuted to the decentralization of more affluent home-owners, and the concentration of the disadvantaged in inner city areas. More recently, substantial reductions in government support for council housing, rising local authority rents, and sales of council housing have increased the concentration of the lowest income groups in council housing and exacerbated the problems of declining inner area neighbourhoods. On the other hand, housing rehabilitation policies and subsidies to encourage home-ownership have had positive effects on the social composition and physical environment of older urban neighbourhoods with spillover effects on surrounding neighbourhoods.

Explicit central government economic development policies have also had important effects. Government post-war policies to redirect manufacturing away from the cities through Industrial Development Certificates, regional incentives, and the development of New Towns affected Birmingham, London, and Bristol. The effects can also be seen in the experiences of the two study areas which have been long-term recipients of a multiplicity of government assistance programmes, Clydeside and Tyneside. This assistance, combined more recently with the substantial expansion of the expenditure of the Manpower Services Commission (MSC), far outweighs local authority expenditure on development. Regional and industrial policies helped to maintain, modernize, and diversify the traditional manufacturing firms of these two areas, and at least in the case of the Newcastle metropolitan area apparently contributed significantly to slowing the decline of manufacturing employment in the 1970s. However, the concentration of these 'reactive' policies on a limited number of traditional capital-intensive manufacturing industries failed to address the fundamental structural problems of the areas and the need to promote new sectors of economic growth. Regional policy, administered mainly by automatic grant support for the plant and equipment investments of firms, failed to support a more strategic and selective approach to the area's problems and actively to promote economic diversification through new firm formation, and the growth of advanced manufacturing industries and business services. Thus, the recent recession has had a devastating effect on a Newcastle metropolitan economy still dominated by traditional manufacturing industries.

Moreover, since regional policy does not have the metropolitan economy or inner city employment problems as its focus, and is largely responsive to the existing geographic distribution of industry and business investment, its effect on either is fortuitous and has varied between areas. For example, with regard to inner city industrial development, regional policy has supported it in the Newcastle case, with approximately two-thirds of assistance from the Department of Trade and Industry going to inner city establishments, and worked against it in Clydeside, where 68 per cent of assistance has gone to the outer areas. The Urban Programme, with its inner city focus, also does not provide an adequate framework for economic development policies to address the structural economic problems of urban regions.

The MSC, with, for instance, annual expenditures in the Newcastle metropolitan area now larger than the combined expenditure of the Department of Trade and Industry and the Department of the Environment, should also play an important role in urban economic development. This is particularly the case in the present period of increasing knowledge-based competition in which skill levels are important to area economic and business growth. However, while the government has accepted the importance of training policies to national economic development, it has not done so with regard to local economic development (McArthur and McGregor 1986). The MSC lacks a metropolitan view of labour markets and their relationship to urban economic adjustment and development, and adequate labour market planning machinery has not been developed for urban areas. The increasing emphasis on demand-led employer-based employment and training schemes may disadvantage urban areas with depressed market economies and inner city residents less attractive to private employers. The MSC has not encouraged urban authorities attempting to develop manpower policies related to local needs and longer-term strategic economic objectives. Lastly, impediments exist to tailoring national manpower policies to local conditions and making innovative use of those policies.

The Urban Programme of the Department of the Environment has promoted economic development efforts by local authorities and increased the investment dedicated to economic and physical development in inner cities; it has also supported some innovative projects and ones with implications for the restructuring of the wider urban economy—for example, university-linked science parks. However, the policy's role has been limited and largely ameliorative. The Newcastle study suggests that the programme's funds are being used largely as another budget line for the existing economic policies of local authorities rather than as an instrument for the development of more effective urban development policies. The West Midlands study argues that a number of factors have restricted inner city policy from being more long-term oriented and innovative: these include regulatory constraints, annual budget cycles, limited resources, and a capital bias in the programme. The Urban Programme has come to be dominated by a short-term process of bidding for projects rather than any strategic considerations. The structure of inner city policy also no longer appears to be a significant source of innovation in urban policy, as central and local government policies have increasingly diverged.

The disparate functional programmes of central government departments have not been effectively co-ordinated in the past or given sufficient flexibility and local administrative latitude to be combined and tailored to distinctive local needs and strategic development objectives. There are still significant organizational, administrative, and policy impediments which need to be overcome to achieve more effective local co-ordination of central government economic policies, despite several recent initiatives such as the establishment of interdepartmental City Action Teams. Moreover, these limited initiatives represent a central government effort independent of the central–local government machinery of inner city policy. An adequate framework for central-local government (or public–private sector) co-operation in long-term urban economic regeneration has not yet been achieved in distressed English urban areas. Moreover, central government urban policy has increasingly moved in the direction of property development, the attraction of private investment, and short-term market opportunities, diverging from local concerns with employment and the disadvantaged. The establishment of urban development corporations independent of local authorities and the provision of direct grant assistance to private urban development schemes are other important elements in the growing separation of local authorities from government urban renewal policies.

On the other hand, the studies also present a mixed picture of the role and influence of local authorities on urban economic change. Urban authorities have responded to the problems of rising unemployment with an expansion of local economic policies. Within the diversity of local authority activity, some general trends emerge. Local efforts have evolved from traditional concerns with land use regulation to the use of land and premises for economic development, direct financial and advisory assistance to firms, and employment and training assistance to disadvantaged urban residents (Mills and Young 1986); however, they appear to have had only modest success in the newer policy areas. Their main efforts still revolve around land and premises, with a strong emphasis on new and small firms. The major local concern has been for employment and distributive issues, and not with the restructuring of the urban economy. The Newcastle study suggests that, while local authorities have given greater attention to small firms and the service sector than has regional policy, local financial assistance to firms is still concentrated on traditional manufacturing industry.

For the most part, local economic policies have been reactive, traditional in character, short term and non-strategic in orientation, *ad hoc*, uncoordinated, modestly financed, and marginal in their effects. Significant impediments exist to local corporate planning and the bending of main programmes in support of economic objectives. The economic capabilities of local government have been limited by constraints of organization, professional expertise, resources, and politics, as well as by central government financial constraints and lack of support for an active local public sector role. The fragmented, committee-based structure of local government combined with its short-term fiscal planning has mitigated against the development of a coherent economic and employment strategy, suggests the West Midlands study. At the same time, the abolition of the English metropolitan counties removed some of the few examples of more ambitious urban economic development policies, and a potential framework and resources for labour-market-wide development policies needed to address urban region economic problems. Moreover, urban authorities are now faced with increasing competition from other local authorities as economic development policies have spread to authorities in non-urban and more prosperous areas (Mills and Young 1986).

However, the research also indicates that local policies have influenced locational decisions and can make positive, important, innovative, and cost-effective contributions to urban economic development and job creation as well as training and anti-poverty initiatives. Local activity is also essential to contemporary development because strategies need to be customized to the local economic situation, comparative advantages, development opportunities, and capabilities. It also requires local entrepreneurial activity to identify and implement development projects and collaborate with the private and voluntary sectors. In the more supportive context of Bristol's locational advantages and better economic performance, the local authority has contributed to growth through its support for development, more flexible planning policies, support for commercial and office development and inward investment, marketing, and the provision of sites and premises. Local authorities such as Bristol have also been expanding their policies to sectors of economic growth such as tourism and advanced technology. In the West Midlands and Greater London, more comprehensive and strategic approaches to metropolitan economic development and the problems of the disadvantaged were introduced by the county councils. These efforts included underpinning action with increased economic analysis, attention to industrial sectors, substantial commitments of financial resources, and the establishment of specialized organizations and instruments in business finance and advice. Birmingham city government has also initiated more aggressive development policies. Evaluations are still needed to assess the effectiveness of these approaches. However, there is evidence to suggest that local authority policies could make a greater contribution to national industrial, area development, and employment objectives.

In Scotland, the existence of a large well-financed professional regional economic development organization providing a variety of specialist services, the Scottish Development Agency (SDA), has substantially assisted longer-term regional economic development and restructuring, as well as urban redevelopment, and has been a valuable complement to local authority efforts while fostering co-operation and a division of labour with local authorities in economic policy. However, there has also been tension between the SDA's objective of regional economic restructuring and its focus on regional development opportunities, and urban authorities' concerns with distressed area regeneration, employment, and social benefits.

Nevertheless, the SDA experience does underscore the importance of several factors in achieving more effective area economic development policies. Its programmes reflect the need for a range of strategies which extend beyond property development (the major element of central government's urban policy) to industrial restructuring, sectoral development, business development, technological innovation and diffusion, and commercial development, based on a firm foundation of economic intelligence. It demonstrates several of the objectives of long-term development policies: to provide the supply-side conditions for economic growth (for example, finance, infrastructure, entrepreneurship, and technical information); to build on the existing comparative advantages of an area and develop new advantages; to attract multiples of private investment and establish public–private sector partnership; and to identify, package, and promote opportunities and catalyse the development process.

The SDA also demonstrates the combination of resources, expertise, and services which is needed to mount economic development policies relevant to the scale of urban and regional economic problems. It highlights the importance of establishing a capability to design and implement effective economic development policies, without which improvements in resources, strategies, and instruments are insufficient. Lastly, it suggests an important role for central government in strengthening urban economic development organizational and professional capabilities at the local level, as well as providing adequate financial resources. At the moment, the organizational arrangements do not exist in the urban areas of England to implement more effective urban region economic development policies. Nor has the role of local authorities in urban development been satisfactorily resolved.

It is apparent that more than traditional development policies are needed to foster economic regeneration in distressed urban regions. There is a need to mount long-term strategies and to support developments which will strengthen the economies of urban regions based on their distinctive problems and impediments to adjustment and growth, their 'indigenous potential', and the new sectors of economic expansion and employment growth and the factors which influence the location and growth of those sectors. The ESRC Programme identifies several necessary elements of more effective economic development strategies: promotion of the expansion of the service sector, particularly producer services; assistance in product and process innovations in local firms, particularly the small and medium-sized firms largely neglected by public policy, in order to enhance their competitiveness and thereby retain and expand them; assistance to advanced-technology firms and industries; strengthening of the factors which help firms to attract technical, professional, and management staff; and development of the skilled labour force necessary to higher-value-added economic activity.

The Bristol study places particular emphasis on the importance of service activities to both the production and service sectors of the economy, the importance of business services, and the importance of white-collar occupations and the service sector as major sources of future employment. With the service sector accounting for two-thirds of total employment and 60 per cent of gross domestic product, and with its output expanding, it needs to be taken seriously by policy-makers. 'The extent to which urban areas attract or generate service growth, and the structure of that activity, will be crucial determinants of economic prosperity and the scale and pattern of employment opportunities' (Boddy *et al.* 1986, 207). The study 'suggests the need for a major shift in policy towards promoting the growth and development of service activities and the service infrastructure' (Boddy *et al.* 1986, 218), and the need to devise new policies to assist services. However, as yet, we have only very limited knowledge of what factors affect the location of producer services and how public policies might intervene to influence the location and growth of this sector. Moreover, the research also suggests that the manufacturing sector should not be discarded as an important element of urban development policies.

With regard to high-technology industries, the city studies offer some cautionary advice. The Bristol study questions the cost-effectiveness of a major local government policy emphasis on advanced technology, given that the potential for 'high-tech' growth is limited and localized and the specific circumstances which favour growth, such as existing concentrations of technologically advanced activity, cannot easily be reproduced. Here again, as in the case of producer services, our knowledge of the factors influencing the location of high-technology industry, and their implications for public policy (for example, defence procurement policies and the location of government R&D facilities) is very limited. Progress on the issue is important in the face of an increasing number of technology-oriented local development initiatives such as science parks and university-based initiatives to promote sectoral and new enterprise development.

The research also suggests that important questions remain with regard to the role of small businesses in employment creation and economic restructuring, and that small-business development policies can only be an element of a comprehensive urban development policy. Lastly, labour market policies must be an important component of urban economic adjustment policies. Skill levels and education are important determinants of urban economic performance, and attitudes and skills are relevant to entrepreneurship. Therefore, investment in an area's human resources through education and training is critical to providing the conditions for economic development.

Public Policy and the Disadvantaged

Alongside the long-term urban policy objective of restructuring, strengthening, and expanding the urban region economy, the ESRC-funded inner cities research underscores the separate and urgent need to address *directly* the short- and long-term employment and training needs of disadvantaged urban residents, to formulate policies 'targeted' on people and not only geographic areas such as inner cities. More effective policies are required to link economic development and employment generation policies with the social objectives of urban policy.

A number of factors necessitate such targeting. Inner city policies, although citing urban deprivation and unemployment as justifications and criteria for the designation of areas for government assistance, have not addressed the issue of who are the beneficiaries of economic development policies. They have assumed that a general rise in local economic activity and employment opportunities will 'trickle down' employment benefits. However, as noted earlier, local employment growth in circumstances of high unemployment in the wider labour market does not lessen significantly the unemployment of those disadvantaged in the competition for employment: there is a weak link between local employment growth and unemployment. In addition, the Newcastle study demonstrates that the new employment resulting from publicly assisted inner city development projects is distributed in much the same way as pre-assisted employment was and as the urban labour market distributes other employment opportunities. Residents outside the target area are the major beneficiaries of new employment, and even among inner city residents employment tends to go to more skilled applicants. Furthermore, the smaller the firm—and small firms are the major targets of inner city economic policies—the more likely the jobs are to go to more skilled individuals. Several new local jobs would have to be created to provide one new job to a disadvantaged urban resident. Moreover, the mobility through migration and commuting of disadvantaged inner city residents to more prosperous localities and regions with more employment opportunities—which would also ease the concentration of deprivation in inner city areas—is hampered by the lack of low-cost rental housing and by transportation costs (Buck and Gordon 1986). However, in the absence of an overall increase in employment opportunities, mobility and other targeting policies will be limited to the social and geographic redistribution of employment.

The research identifies a number of other problems with economic policies as they affect the employment of the disadvantaged. Employment in small businesses, which are the focus of much of local development policy, is skewed towards more skilled workers as compared with the higher proportion of less skilled workers in the larger manufacturing establishments which have been the focus of regional policy. Conventional market-orientated economic development programmes are difficult to implement in severely distressed communities where the disadvantaged reside, because of the lack of development opportunities. At the same time, the Urban Programme and the policies of the MSC restrict the development of community businesses, an alternative employment source.

Manpower policies present several other impediments to assisting the disadvantaged (McArthur and McGregor 1986). The increasing emphasis on employer-based and market-orientated employment and training schemes limits the resources directed to distressed urban economies with less demand and the training given to the disadvantaged. Cost-effectiveness criteria reinforce an aversion to addressing the costlier needs of disadvantaged clients and areas. Self-employment programmes have not been taken up as well in distressed areas and areas dominated by large manufacturing industries without an entrepreneurial tradition. Temporary employment and self-employment programmes have lacked training programmes for the disadvantaged and long-term unemployed. More generally, constraints on local public expenditures have eliminated stable employment for the less-skilled, while encouragement of low-paying service sector employment threatens to consign them to casual dead-end jobs (Buck *et al.* 1986).

New Directions in Urban Economic Policy

The perceptions of urban problems have changed substantially over the last twenty-five years, from an initial concern with urban congestion and physical decay to concerns about concentrations of multiply deprived people in inner city areas, to the overall economic and social well-being of older industrial urban areas affected by counter-urbanization and de-industrialization. Urban policy in the late 1970s recognized the problem of cities as a whole suffering from major structural changes, and governments have responded with greater priority for the more distressed areas and increased emphasis on environmental, property, and economic redevelopment, additional resources, a call for 'bending main programmes' in support of urban policy objectives, efforts to attract private sector investment, and innovations in instruments and agencies—although still with a focus on inner city areas. In addition, and partly with central government support, urban authorities have increased their economic activities in the face of very high unemployment. However, the research indicates that public policy has still failed to articulate adequately, and to distinguish among, its policy objectives, and to address adequately in the scale or character of its efforts the dual problems of urban region-wide economic decline and the concentration of unemployment, poverty, and social deprivation in inner city areas and peripheral estates. These are separate, although related, problems which require different policy responses.

Major processes of structural change have altered the prospects for urban economic adjustment and the requirements for future urban development, and have redefined the policy choices open to public officials. Renewed national economic growth, while a necessary condition for addressing the problems of disinvestment and unemployment in cities, will not ensure their solution while urban impediments to adjustment and development remain. Urban job creation will not ensure employment for inner city residents who are disadvantaged in urban labour markets. Urban economic policy must encompass three distinct objectives:

(*a*) The long-term restructuring and development of the economies of the old industrial urban regions
(*b*) Increasing the employability and employment of inner city residents
(*c*) Amelioration over the shorter term of the fiscal problems of urban auth-

orities and the social and employment problems of disadvantaged urban residents who bear the brunt of economic change.

The issue is not just one of managing the urban and regional decline which has already struck with cataclysmic force, but rather of fostering the economic and social adjustment and revitalization of urban regions which can provide cities and their residents with the better of the alternative economic and social futures open to them in the new national and international circumstances. In the few remaining pages I will outline the framework of policies which the research suggests to me are necessary to respond effectively to urban economic and employment problems.

Despite the recognition in the late 1970s of the economic underpinnings of urban decline, the objective of *urban economic restructuring* has not been adequately addressed by government policy. The research suggests some of the elements of a more substantial approach to economic redevelopment. The policies must be long term, strategic, and focused on the wider urban economy and its region, and not just on the inner city. Economic restructuring needs to extend beyond the preservation and modernization of the traditional economic structure, modest *ad hoc* employment creation projects, and property development schemes, to promote new sectors of self-sustaining economic activity with better employment, income, and wealth creation potential. Restructuring policies involve strategies to promote key sectors of the economy (including, but not exclusively, manufacturing), to assist the development of new *and* established businesses, to stimulate area development programmes, and to establish the supply-side conditions (for example, skills, land and premises, infrastructure, entrepreneurship, information, attitudes, environment, housing, and amenities) at competitive costs to support economic adjustment and development. They must address the restructuring process of corporations to try to influence their investment and location decisions. They must build on existing comparative advantages ('indigenous potential') as well as investing in the establishment of new competitive advantages (that is, the concept of 'dynamic comparative advantage': Scott 1984)—for example, through support of centres to technical research in key technologies working with industry. The instruments of such policies include (in addition to land and premises) direct financial and advisory assistance to businesses and industries, the diffusion of technical information and economic intelligence, and innovative methods of using these instruments. In addition, they must be supported by the co-ordinated use of local main programmes and the levering of private investment.

With regard to economic restructuring, local economic policies must be undertaken in the context of concerted action by central government. Central government must make a number of essential contributions to strategic urban development policies:

(*a*) Provide a context of macro- and micro-economic policies which stimulate national economic growth and business investment

(*b*) Provide information on the geographic distribution of government expenditures and assessments of their area development impacts
(*c*) Use selected policies (for example, R&D and defence procurement) to support industrial and area development objectives
(*d*) Co-ordinate economic policies (for example, industrial, regional, urban, and manpower) in support of urban development objectives
(*e*) Dedicate resources over the long term to area development programmes and strategies, and not just to the financing of *ad hoc* projects
(*f*) Allow flexibility in the use of government programmes and decentralization of their administration to facilitate the tailoring of programmes to local needs and to encourage policy innovation
(*g*) Provide support for building the capabilities of area development agencies to design and implement more effective strategies
(*h*) Provide comprehensive and timely economic information to support strategic area development policies
(*i*) Provide adequate resources for area development, relax constraints on the use of local resources for development purposes, and establish adequate incentives to stimulate private sector reinvestment in distressed urban regions.

Lastly, effective restructuring and development policies will require professional area development agencies, combining public and private sector skills, to catalyse and facilitate the development process working co-operatively with the public, private, and voluntary sectors. The experiences of the regional development agencies, local authority development organizations, urban development and new town development corporations, enterprise boards, and private sector development organizations should be examined in the development of effective delivery systems for urban and regional development policies.

Secondly, urban policy must include separate *'targeted' employment policies* that are concerned with the social distribution of employment and the competitiveness of groups and individuals in the labour market, and which provide direct assistance to disadvantaged urban residents. The purpose of such policies would be to remove particular barriers to labour market adjustments and enhance the employability of, and the employment opportunities for, those less competitive in the labour market. They are based on the recognition that specific practical actions are necessary to forge links between the disadvantaged and the urban economy and employment. Effective outreach to inner city residents will be necessary to involve them in such efforts. Effective action to reduce concentrated deprivation and unemployment is also an important condition for urban economic revitalization by improving the labour force of the area and removing an important disincentive to private investment.

A number of measures should be considered in formulating such targeted policies. Long-term education and training programmes must be focused on the disadvantaged in order to improve their competitiveness, even if this requires higher per caput training costs and support for special training arrangements less dependent on local market conditions and private sector initiatives. Policies to improve the general 'suitability' of certain inner city groups to private sector recruiters, to encourage greater receptiveness by the private sector to the recruitment of specific urban groups, and to combat any discriminatory practices are important. Policies which would improve the employment opportunities available to disadvantaged urban residents include:

(*a*) Job preservation strategies aimed at assisting businesses already employing target urban groups and retraining their workers to meet new demands
(*b*) More effective placement programmes to bring selected individuals, including graduates of public sector training programmes, to the attention of private sector and other employers
(*c*) Linking of public sector training and placement programmes with publicly assisted business development, construction, and employment creation projects in order to target the resulting employment opportunities (this will include employing local labour in environmental, housing, and other community improvement schemes in distressed areas of the inner city and peripheral estates)
(*d*) Development assistance to sectors and firms which are more likely to employ the less-skilled and those without formal qualifications
(*e*) The targeting of self-employment and new enterprise development schemes on the disadvantaged
(*f*) Support for community enterprises and other special employment creation measures, although such measures may only be modest in their effects
(*g*) Incentives to business to employ the disadvantaged and unemployed—for example, publicly financed training of disadvantaged individuals tailored to specific private employers, and wage subsidies.

The mitigation of residential segregation would also contribute to reducing the disparity in unemployment rates evident in inner city areas. This would entail the removal of housing, planning, transportation, and other impediments to the mobility of inner city residents, particularly in the more prosperous south of England. Increasing the opportunities for out-migration from the inner cities would also lessen the negative effects of concentrated deprivation in inner city areas. In developing targeted employment policies the experiences of the variety of existing innovative local labour market and community development policies need to be assessed. Targeted labour market policies will also require effective delivery systems and organizations capable of labour market planning, designing and implementing programmes, and mobilizing community support and participation, as well as the involvement of the private and voluntary sectors. The potential of community development agencies, housing associations, enterprise agencies, and other community and voluntary organizations to play a more important role in community economic policies needs to be explored.

Alongside these longer-term development and labour market policies, a third policy strand is required: *amelioration of the consequences of urban economic change*. The adjustment to major structural and social change has caused significant problems for people, businesses, and local government which must be addressed in the short term, because they will not be quickly alleviated by restructuring of the private market economy, the generation of private sector employment, and the strengthening of the local tax base. In circumstances of persistent high unemployment, consideration should be given to the provision of alternative employment opportunities. These include the creation of publicly financed long-term employment opportunities: for example, employment on labour-intensive environment, housing, infrastructure, and social service projects; the development of special enterprises; and employment subsidies. Moreover, social services, community development, housing, education, and other assistance to the poor and unemployed should not be neglected through excessive concentration on economic development and employment measures. Commercially viable urban businesses, faced with the loss of large corporate clients, increased competition, lower demand, and the need for costly adjustment of their operations, could be assisted during the difficult adjustment period in order to increase their chances of survival. Similarly, major industrial corporations in older industrial areas undertaking measures of rationalization, restructuring, and diversification could be assisted during this period in order to encourage their adjustment in place. For local authorities in distressed urban areas, central government constraints on local finance and reductions in financial assistance could be eased in order to maintain needed local services and employment, to enhance the area's attractiveness to business investment and residents, thereby slowing the spiral of urban decline, and to increase local development efforts.

The fourth component of urban policy is *capacity-building*, strengthening the capability to develop and implement programmes and projects to achieve economic restructuring and development, labour market adjustment, and the alleviation of distress. There are a number of facets to the policy of 'capacity-building':

(*a*) Clearer articulation of urban policy objectives and evaluative criteria
(*b*) The development of effective machinery for the co-ordination of central government main programmes and their flexible, innovative, and tailored response to urban needs
(*c*) The improvement of central government programmes and instruments, and the adequate financing of them

(*d*) The development of effective machinery for central–local government co-operation in long-term strategic urban economic adjustment and development

(*e*) Development of a positive role for local authorities in urban development; the provision of technical, training, and financial assistance to local authorities and the lifting of harmful constraints in order to enhance local capability, encourage a longer-term development perspective and the bending of main programmes, and provide an incentive for local economic policies

(*f*) Promotion of the establishment of effective machinery for private sector investment and participation in broader urban redevelopment efforts, and public–private sector co-operation

(*g*) Strengthening of development organizations and professional capabilities necessary for community economic development and addressing the problems of the disadvantaged

(*h*) Establishment and strengthening of specialized development agencies and finance organizations in urban regions with adequate professional expertise and resources to address area-wide strategic economic development objectives

(*i*) Provision of technical assistance and information to local authorities and community and other organizations in order to improve the take-up and use of central government programmes.

Lastly, more strategic and effective urban and regional economic policies will require better *economic intelligence*. There is a need for more comprehensive, regular, accessible, and in-depth national, regional, and local information on economic and labour market conditions and trends. Information and analyses of the patterns and area effects of central and local government general expenditures is important to make public sector policy more sensitive and responsive to urban development objectives. In addition, assessments of the effectiveness of particular policies, programmes, projects, instruments, organizational arrangements, agencies, and innovative activities—and diffusion of information on 'best practice'—are essential to improving the performance of urban economic policies.

A number of general considerations apply to the formulation of an overall urban economic policy. The policy will include a range of action from providing a supportive climate for business investment to more focused intervention. It will require more entrepreneurial activity by the public sector in implementing development projects, collaborating with the private sector, and seeking a return (that is, employment, private investment, and tax revenues) on the investment of public resources. It will have to balance economic and social objectives. A variety of public sector activities relevant to the development process will need to be mobilized. Public resources must aim to lever private investment. The professional and organizational framework for the effective delivery of policy and collaboration with the private sector and the community will have to be established. Public policies will need to be monitored and evaluated to determine their effectiveness in achieving specifically defined operational objectives. Urban economic redevelopment is a long-term process which will require continued political and public support at the national and local levels.

The five city studies and the other research of the ESRC Inner Cities Research Programme have demonstrated that structural change is having profound effects on the British urban system, urban economies, and labour markets, and on the geographic and social distribution of employment opportunities. They have also identified significant limitations on current policies to address effectively both the economic adjustment problems of urban regions and the employment needs of disadvantaged urban residents. Finally, they have suggested a number of the policy elements of a more effective approach to these problems. Those readers who wish to explore the city study findings in greater depth and the ESRC Programme's other research findings will find individual volumes on each city study and two volumes of reports on special research projects as part of the publication series of the Inner Cities Research Programme.

Notes

1. The six large conurbations are: London, West Midlands, Greater Manchester, Merseyside, Tyneside, Clydeside.
2. The seventeen large free-standing cities are: Sheffield, Leeds, Edinburgh, Bristol, Coventry, Nottingham, Bradford, Hull, Leicester, Cardiff, Stoke, Plymouth, Derby, Southampton, Aberdeen, Portsmouth, Dundee.
3. Source: *Social Trends*, 1984.

References

Begg, I., and Eversley, D. (1986), 'Deprivation in the Inner City: Social Indicators from the 1981 Census', *Critical Issues in Urban Economic Development*, Vol. i, Oxford: Clarendon Press.

—— and Moore, B. (1986), 'The Changing Economic Role of Britain's Cities', in V. A. Hausner (ed.), *Critical Issues in Urban Economic Development,* Vol. ii, Oxford: Clarendon Press.

—— —— and Rhodes, J. (1986), 'Economic and Social Change in Urban Britain and the Inner Cities', in V. A. Hausner (ed.), *Critical Issues in Urban Economic Development,* Vol. i, Oxford: Clarendon Press.

Boddy, M., Lovering, J., and Bassett, K. (1986), *Sunbelt City? A Study of Economic Change in Britain's M4 Growth Corridor*, Oxford: Clarendon Press.

Breheny, M., Cheshire, P., and Langridge R. (1983), 'The Anatomy of Job Creation? Industrial Change in Britain's M4 Corridor', *Built Environment*, 9(1), 61–9.

Buck, N., and Gordon, I. (1986), 'The Beneficiaries of Employment Growth: An

Analysis of the Experience of Disadvantaged Groups in Expanding Labour Markets', in V. A. Hausner (ed.), *Critical Issues in Urban Economic Development*, Vol. ii, Oxford: Clarendon Press.

—— —— and Young, K. (1986), *The London Employment Problem*, Oxford: Clarendon Press.

Champion, A.G., and Green, A.E. (1985), 'In Search of Britain's Booming Towns: An Index of Local Economic Performance for Britain', *Discussion Paper* 2, Centre for Urban and Regional Development Studies, University of Newcastle.

Ermisch, J., and Maclennan, D. (1986), 'Housing Policies, Markets, and Urban Economic Change', in V. A. Hausner (ed.), *Critical Issues in Urban Economic Development*, Vol. ii, Oxford: Clarendon Press.

Hall, P. (1981), *The Inner City in Context*, London: Heinemann Educational Books.

—— (1985), 'Restructuring of Urban Economies: Integrating Urban and Sectoral Policies', Paper prepared for OECD Seminar on Opportunities for Urban Economic Development, Venice, Italy, June 1985.

Kirwan, R. (1986), 'Local Fiscal Policy and Inner City Economic Development', in V. A. Hausner (ed.), *Critical Issues in Urban Economic Development*, Vol. i, Oxford: Clarendon Press.

McArthur, A.A. and McGregor, A. (1986), 'Local Employment and Training Initiatives in the National Manpower Policy Context', in V. A. Hausner (ed.), *Critical Issues in Urban Economic Development*, Vol. ii, Oxford: Clarendon Press.

Markusen, A. (1984), 'Defence Spending and the Geography of High Tech Industries', *Working Paper* No. 423, University of California, Institute of Urban and Regional Development, Berkeley.

—— (1985), 'The Military Remapping of the United States', *Built Environment*, 11(3), 171–80.

Mills, L. and Young, K. (1986), 'Local Authorities and Economic Development: A Preliminary Analysis', in V. A. Hausner (ed.), *Critical Issues in Urban Economic Development*, Vol. i, Oxford: Clarendon Press.

Noyelle, T.J., and Stanback, J.T.M. (1984), *The Economic Transformation of American Cities*, Totowa, NJ: Rowman and Allanheld.

Saxenian, A. (1983), 'The Genesis of Silicon Valley', *Built Environment*, 9, 7–17; reprinted in P. Hall and A. Markusen (eds.), *Silicon Landscapes*, London: Allen and Unwin, 20–34.

Scott, B.R. (1984), 'National Strategy for Stronger US Competitiveness', *Harvard Business Review*, 62 (Mar.–Apr.), 77–91.

Smith, N. and Williams, P. (eds.) (1986), *Gentrification of the City*, Boston, Mass.: Allen and Unwin.

Steiner, M. (1985), 'Old Industrial Areas: A Theoretical Approach', *Urban Studies*, 22, 387–98.

Wolman, H. (1985), 'Urban Economic Performance', unpublished paper prepared for ESRC Inner Cities Research Programme.

—— (1986), 'Urban Economic Performance: A Comparative Analysis', in V. A. Hausner (ed.), *Critical Issues in Urban Economic Development*, Vol. ii, Oxford: Clarendon Press.

2

Bristol: Sunbelt City?

*Martin Boddy**

The easy image of Bristol is that of a city successfully restructuring in the face of national recession and rising unemployment—part of the M4 growth corridor, 'Britain's Sunrise Strip',[1] on which national hopes of economic renaissance have been pinned. It was, according to the *Sunday Times* in 1981, 'well on the way to becoming the high technology centre of Britain . . . set to become Britain's Silicon Valley of the 1980s'.[2] Unemployment had of course risen as elsewhere: by early 1987 it was approaching 11 per cent. This was, nevertheless, significantly down on the national figure of over 13 per cent and compared very favourably with Glasgow, Newcastle, or the West Midlands. Government indices of 'urban deprivation' suggest that the area as a whole shares little of the problems of London's inner boroughs or the cities of the Midlands and the North. Reinforcing this image, 'high-tech' firms including Hewlett Packard, Digital, and Inmos have set up in the city. Major insurance companies have relocated their headquarters operations to the expanding office-based service sector. Shrinking employment in traditional manufacturing has largely been compensated for, numerically at least, by the growth of service employment and the continued prosperity of technologically advanced manufacturing, particularly aerospace. And all this has been achieved without the range of explicit urban and regional policy measures directed to the regions and inner cities.

As these examples suggest, however, success has been underlaid by major shifts in economic structure. This in turn has impacted on the labour market and the fortunes of different groups in the local population, both in and out of formal employment. As elsewhere, male manufacturing employment has been replaced by female service employment, much of it part time, with the sunrise industries doing little to compensate for the collapse of much of the manufacturing sector. Youth and long-term unemployment rates have risen towards national levels and there are pockets of high unemployment and multiple deprivation in the inner area and in outlying council estates that are as severe as those in the conurbations of the Midlands and the

* Martin Boddy was responsible for producing this particular account. However, the chapter very much reflects the joint work of the Bristol project team, in particular John Lovering and Keith Bassett. Frankie Ashton and Tom Davies also made significant contributions to the earlier stages of the project.

North. Moreover, it was in Bristol, not Brixton, that the urban violence of the early 1980s first erupted. While not denying the contrasts with these other localities, it is clearly necessary to look a little more closely at the image of success and effective 'adaptation'.

First, then, the Bristol experience raises the question of what we can learn about the anatomy of success. We need both to focus on the underlying processes and to look more closely at the extent to which the impacts of 'success' reverberate through the structure of the economy and local labour market. Who shares in the benefits of success? How far is public policy, central or local, implicated in the particular patterns of economic and employment change? Are there conclusions to be drawn which may be relevant to the problems of Glasgow, the North East, the West Midlands, and elsewhere? The city is thus, at one level, a specific case-study of a relatively prosperous subregion to set alongside the experience of other localities. As is now increasingly recognized, urban and regional policy analysis has been preoccupied with the 'problem areas' themselves. The processes of change in relatively prosperous localities, intimately related to the same patterns of restructuring affecting the regions and inner cities, have been relatively neglected. However, the Bristol case also exemplifies, in a specific urban context, changes in national and indeed international economic structures. In many ways, it appears to be at the forefront of emerging trends in terms of economic restructuring and labour market change, including the growth of high-technology industry, 'de-industrialization', the expansion of the service sector, and the rise in female employment.

This chapter therefore starts off in the next three sections by looking in turn at economic structure and change, employment and unemployment, and spatial inequalities. It then focuses on a range of policy issues, before concluding with a discussion which draws out some of the wider implications of the analysis as a whole. The analysis is based on material assembled in the course of the Bristol component of the Economic and Social Research Council (ESRC) Inner Cities Research Programme. This included an extensive programme of interviews with a panel of employers in key sectors in the locality (Figure 2.1).[3] This was additional to the wide range of other material, including official statistics, documentary sources, and interviews and discussions with representatives of local authorities, labour market agencies, and other bodies on which the work as a whole was based. Much of this has been written up in more detail elsewhere and is referred to where appropriate.[4] The aim here is to focus on key components of economic structure and employment change, the role of public policy, and some of the wider implications.

Economic Structure and Change

To give an initial picture of economic structure, employment in the locality totalled almost 300 000 in 1981. Manufacturing employment represented 28

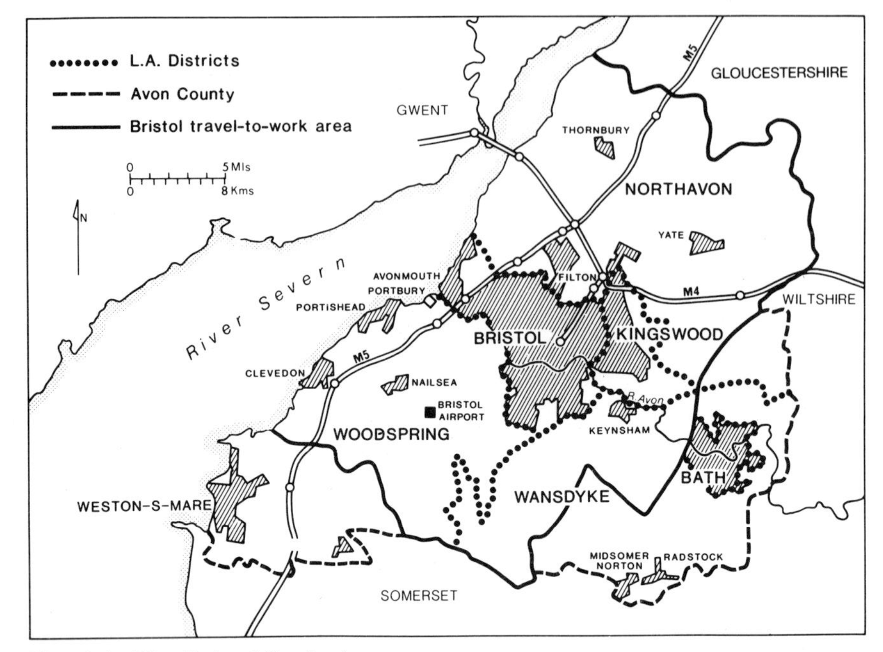

Fig. 2.1. The Bristol Study Area

per cent of the total, roughly the same as nationally, while service employment at 63 per cent was marginally higher.[5] Male employment was slightly more important, and female slightly less, than nationally. The area had a higher than average share of employment in Food, drink, and tobacco, Vehicles (mainly aerospace), and Paper, printing, and publishing in the manufacturing sector; and, in the services, in Insurance, banking, and finance, Professional and scientific services (mainly education, medical, and dental employment), Transport and communications, and Gas, electricity, and water (Table 2.1). The main employment sectors are broken down in more detail in Table 2.2. This emphasizes the importance in terms of total employment of a range of activities (particularly in the service sector), such as education and retailing, which in percentage terms are no more important locally than nationally.

Employment fell by under 1 per cent (2500) over the decade to 1981. This, however, disguises more fundamental changes. First, manufacturing employment fell by over 23 500, largely offset by the rise in service jobs. Second, male employment dropped almost 16 000, largely offset, numerically at least, by a 13 500 increase in female employment (Table 2.3). The major increase, in fact, was specifically in female service employment, much of it part-time work. Miscellaneous services employment, for example, increased by nearly 2400 from 1978 to 1981. Female employment, however, accounted for 2000 of this, and part-time female employment for nearly 1600. Female manufacturing

Table 2.1. *Employment by Sector, Bristol and Great Britain, 1981*

SIC 1968		Bristol				Great Britain (%)	Location quotient[a]
		Male	Female	Total	Total (%)		
I	Agriculture, forestry, fishing	2 027	1 472	3 499	1.17	1.75	0.67
II	Mining and quarrying	495	78	573	0.19	1.58	0.12
I–II	All primary	2 522	1 550	4 072	1.36	3.33	0.41
III	Food, drink, and tobacco	9 309	6 362	15 671	5.26	3.00	1.75
IV	Coal and petroleum products	24	18	42	0.01	0.13	0.08
V	Chemicals and allied industries	4 398	704	5 102	1.71	1.90	0.90
VI	Metal manufacture	1 135	213	1 348	0.45	1.49	0.30
VII	Mechanical engineering	7 323	1 111	8 484	2.83	3.63	0.78
VIII	Instrument engineering	593	292	885	0.04	0.63	0.06
IX	Electrical engineering	3 334	1 060	4 394	1.47	3.18	0.46
X	Shipbuilding and marine engineering	284	19	303	0.10	0.68	0.15
XI	Vehicles	22 768	2 672	25 440	8.53	2.79	3.06
XII	Metal goods n.e.s.	2 932	736	3 668	1.23	2.10	0.59
XIII	Textiles	278	176	454	0.15	1.49	0.10
XIV	Leather, leather goods, and fur	143	30	173	0.06	0.14	0.43
XV	Clothing and footwear	604	1 122	1 726	0.58	1.25	0.46
XVI	Bricks, pottery, glass, cement, etc.	1 092	187	1 279	0.43	1.01	0.43
XVII	Timber, furniture, etc.	2 752	528	3 280	1.10	1.02	1.08
XVIII	Paper, printing, and publishing	6 345	2 651	8 996	3.02	2.40	1.23
XIX	Other manufacturing industries	1 879	686	2 565	0.86	1.18	0.73
III–XIX	All manufacturing	65 193	18 567	83 760	28.10	28.01	1.00
XX	Construction	14 650	1 648	16 298	5.42	5.15	1.05
XXI	Gas, electricity, and water	5 118	1 567	6 685	2.24	1.60	1.40
XXII	Transport and communication	18 495	4 496	22 991	7.71	6.71	1.15
XXIII	Distributive trades	19 411	20 529	39 940	13.40	12.85	1.05
XXIV	Insurance, banking, finance, etc.	10 528	11 846	22 374	7.30	6.19	1.18
XXV	Professional and scientific services	16 701	35 433	52 134	17.49	12.85	1.36
XXVI	Miscellaneous services	13 684	21 435	35 119	11.78	11.96	0.99
XXVII	Public administration and defence	9 492	5 271	14 763	4.95	7.17	0.69
XXII–XXVII	All services	88 311	99 010	187 321	62.83	57.73	1.09
I–XXVII	All industries	175 794	122 342	298 136	100.00	100.00	

[a] Location quotient = (percentage share Bristol)/(percentage share GB).

Source: Annual Census of Employment, Department of Employment.

Table 2.2. *Main Employment Sectors, Bristol 1981*

MLH	Activity	% of total employment		Location quotient
		Bristol	GB	
872	Educational services	7.8	8.00	0.98
383	Aerospace	7.3	0.88	8.25
874	Medical and dental services	7.2	6.55	1.11
821	Retail distribution: other	5.2	5.89	0.89
899	Other services	3.4	3.01	1.14
906	Local government service	2.7	4.39	0.62
708	Postal services and telecommunications	2.6	2.02	1.29
860	Insurance	2.6	1.39	1.86
820	Retail distribution: Food and drink	2.4	2.91	0.84
901	National government services	2.3	2.78	0.83
865	Other business services	1.9	1.44	1.32
861	Banking and bill discounting	1.8	1.73	1.01
810	Wholesale distribution of food and drink	1.9	1.06	1.77
812	Other wholesale distribution	1.6	1.26	1.26
482	Packaging products	1.5	0.32	4.53

Source: Annual Census of Employment, Department of Employment.

employment, on the other hand, fell by almost a third. The shift from manufacturing to service employment and from male to female obviously mirrors national trends, but with an exaggerated increase in female service employment, a smaller drop in male manufacturing employment, and a weaker increase in male service employment than nationally. The shift from manufacturing to services thus had considerably less impact on male employment than on female, compared to the national picture.

The pattern of change

Total employment in the locality had grown rapidly through the 'long boom' of the 1950s and 1960s. Employment trends mirrored the national picture, but with Bristol gaining more than its share of national growth when jobs were expanding and losing less in periods of decline. Food, drink, and tobacco, and Paper, printing, and packaging, in particular, were sustained by the prosperity of mass consumer markets, while a succession of major civil and military projects underpinned the aerospace sector. Together, these sustained other elements of the economy including engineering and metal manufacture. Then, as manufacturing growth slowed through the 1960s, services expanded strongly, particularly public employment in education and medicine. Unlike many localities where the post-war picture was of accelerating decline in their traditionally basic industries, it was not really until the

Table 2.3. *Manufacturing and Service Employment Change by Gender, 1971–1981, Bristol and Great Britain*

	Bristol (No.)	Bristol (%)	Great Britain (%)
Total	−2 084	−0.7	−2.3
Manufacturing[a]	−23 660	−22.0	−24.9
Service	21 694	13.1	15.3
Male	−15 604	−8.1	−6.0
Female	13 520	12.4	3.1
Male manufacturing	−14 962	−18.7	−23.0
Male service	1 175	1.3	15.2
Female manufacturing	−8 700	−31.9	−29.1
Female service	20 520	26.2	15.3

[a] Manufacturing, SICs III–XIX; service, SICs XXII–XXVII.

Source: Annual Census of Employment, Department of Employment.

1970s that major job loss and restructuring reverberated through Bristol's dominant manufacturing sectors. Bristol's longer-term legacy had been in industries set to expand in the post-war boom rather than in, for example, coal, iron and steel, shipbuilding, and heavy engineering. From the early 1970s, however, manufacturing job loss accelerated, with the failure of the Concorde project and the impact through the decade of declining demand for the locality's consumer-oriented industries, exacerbated by deepening recession from the late 1970s. While public services growth slowed through the 1970s, private services expanded rapidly, particularly early on. Yet, despite this expansion, there was a sharp drop in overall employment in the early 1970s which brought a brief localized economic crisis that pushed unemployment above the national rate. As the recession deepened, however, and Conservative economic policies started to take effect from the late 1970s, although unemployment continued to rise, the area recovered relative to the national picture.

Major job losses in absolute terms over the decade to 1981 were thus concentrated in the traditionally dominant manufacturing sectors. Food, drink, and tobacco and Printing, paper, and publishing both declined much faster than nationally, losing over 16 000 jobs up to 1981 and several thousand more by 1985 (Figure 2.2 and Table 2.4). There were significant losses also in parts of Aerospace, the Distributive trades, and Metal manufacture. The picture of employment collapse in traditional manufacturing sectors—as rapid in Bristol over the 1970s as in many other localities—countered by the expansion of service employment is familiar enough. However, Aerospace, employing nearly 24 000 in 1971, contracted much less rapidly than manufacturing overall over the decade, and in fact expanded towards the end of

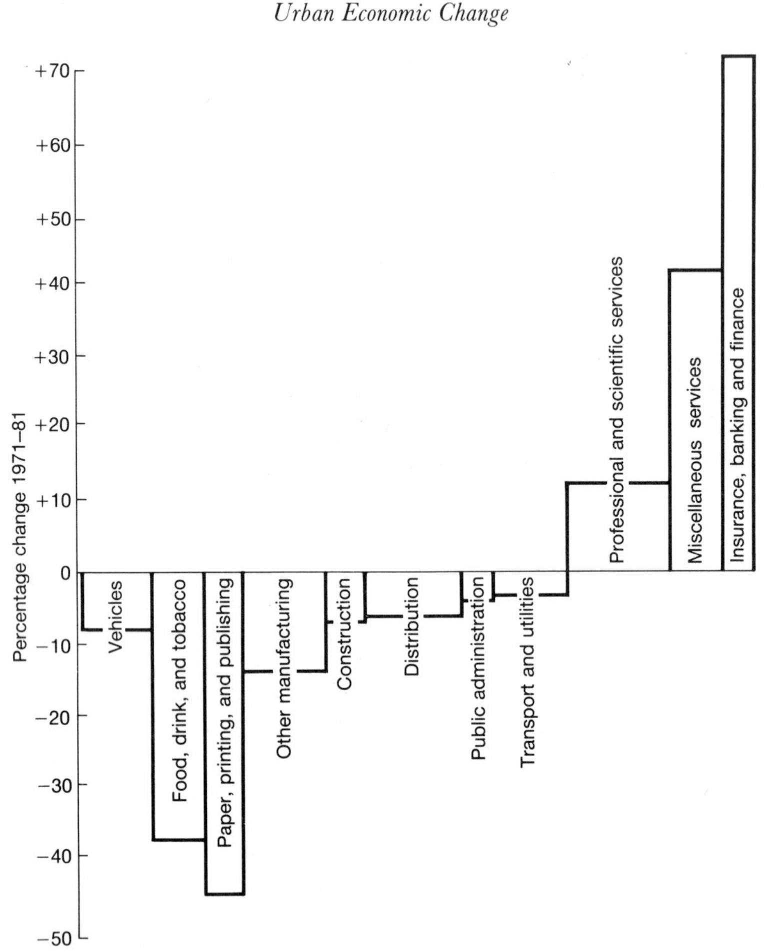

Fig. 2.2. Employment Change in Bristol by Sector, 1971–1981

Note: The width of each sectoral block indicates the proportion of total employment in that sector; the total width of the blocks equals 100 per cent.

Source: Annual Census of Employment, Department of Employment.

this period. This attenuated the fall in manufacturing employment as a whole. And, though there were major losses in other manufacturing sectors, collapse was far from total. In a sense, it took the form of corporate restructuring, reorientation of markets, and 'shake-out' of employment delayed by the continuing boom of the 1960s and 1970s and precipitated by the sharp recession in demand in the late 1970s which brought severe short-term crises for a number of employers. The locality thus shared in the more general failure of manufacturing industry nationally to reinvest and restructure through the long boom (Gamble 1981). While manufacturing employment in general declined, output in many sectors, including aerospace, electronics, and

Table 2.4. *Major Employment Gains and Losses, Bristol 1971–1981*[a]

SIC 1968		Change, 1971–1981 (No.)	(%)
III	Food, drink, and tobacco	−9 000	−36.5
XVIII	Paper, printing, and publishing	−7 342	−44.9
XXIII	Distributive trades	−2 625	−6.1
XI	Vehicles	−2 099	−7.6
VI	Metal manufacture	−2 078	−60.6
XXVI	Miscellaneous services	+10 524	+42.8
XXIV	Insurance, banking, and finance, etc.	+9 440	+73.0
XXV	Professional and scientific services	+5 904	+12.8

[a] Gains and losses greater than 2 000 in any SIC category.

Source: Annual Census of Employment, Department of Employment.

tobacco, continued to expand, even after the late 1970s, emphasizing the fact that job loss is by no means synonymous with economic decline. Aerospace locally, in particular, epitomized the phenomenon of 'jobless growth'.

The major growth was in service employment. There were, however, major contrasts between different components. Much of the growth was in private services. Insurance, banking, and finance expanded at twice the national rate, and Miscellaneous services also outpaced national growth. Professional and scientific services (mainly public sector education, medical, and dental employment) were important in absolute terms, but in fact expanded more slowly than in the country as a whole. Employment in these three sectors together increased by over 25 000 in the decade up to 1981. Employment declined, however, across a broad band of services including transport and communications, wholesale and retail distribution, public administration, and defence, and in the public utilities, Gas, water, and electricity. Further analysis indicates that, from 1971 to 1978, private services expanded by 21 per cent compared with a 7 per cent increase in public services. Then, from 1978 to 1981, as losses in retailing and distribution starteed to offset continuing gains in financial and miscellaneous services, private services increased by only 1 per cent. Public services, however, fell by nearly 5 per cent as cuts in education and government employment outweighed continuing expansion in the medical and dental sector.[6]

Analysis of the area's industrial structure as a whole suggests that a favourable overall mix of industries, biased towards those which performed relatively well in employment terms nationally, underpinned local employment (Table 2.5). This was particularly the case for manufacturing industry where, the analysis suggests, a relatively favourable local industrial structure at the start of the period accounts for the slower decline in manufacturing as a whole compared to the national picture. Services, on the other hand,

Table 2.5. *Shift-Share Analysis of Employment Change,*[a] *Bristol 1971–1981*

	National	Structural	Differential	Actual
Manufacturing[b]	−27 207	+4 036	−494	−23 665
Service	+23 882	+860	−3 050	+21 692
Total	−6 955	+5 668	−1 181	−2 468

[a] Shift-share analysis divides actual employment change into three elements. The *national* shift is that change in employment which would have occurred if employment in each industry had changed at the same rate as total employment in the country as a whole. The *structural* shift is that change in employment, additional to the national shift, which would have occurred if employment in each industry locally had changed at the same rate as in that industry in the country as a whole—an indicator of whether the locality had a favourable mix of industries. The *differential* shift is that change in employment which is additional to the national and structural shifts, indicating the extent to which industries locally have grown or declined faster than those industries nationally. Industries were defined in SIC terms and the results aggregated into manufacturing, services, and total employment.
[b] Manufacturing, SICs III–XIX; Services, SICs XXII–XXVII; Total, Manufacturing + Services + SICs XX and XXI.

Source: Based on Annual Census of Employment, Department of Employment.

expanded marginally less than would have been expected on the basis of national trends and the initial mix of service activities. We can flesh out this basic picture by looking in more detail at a number of particular sectors.

Tobacco Decline in tobacco employment accounts for a significant proportion of the overall drop in the Food, drink and tobacco sector. Employment is concentrated in what is now the Bristol-based Imperial Tobacco Limited, a company within the Imperial Group, taken over by the Hanson Trust in 1986. Imperial Tobacco has production sites in Glasgow, Nottingham, and elsewhere in Great Britain. Employment locally fell from nearly 8000 in 1971 to 5500 by 1981, a drop of over 30 per cent compared with only 19 per cent nationally. Demand for tobacco products fell rapidly from the early 1970s, reflecting the health issue, the decreasing popularity of smoking, and price increases due to sharply increased taxation. Additionally, the company was losing its market share in a fiercely competitive market, bringing strong pressure to increase productivity. Job loss locally through the 1970s reflected capacity cuts plus major new investment in a centralized production and administrative complex. A subsequent programme of investment in new production equipment throughout the tobacco division was overtaken, following a slump in overall group profits in the early 1980s, by a major programme of rationalization, closures, and management reorganization throughout the company's production sites. Production capacity in Bristol was cut (including the closure of a cigar factory), and employment was lost through new investment and rationalization. However, management and administrative functions for the tobacco division as a whole—previously split between Bristol and other sites, particularly Nottingham—were centralized at Bristol.

Bristol's position in the corporate hierarchy was thus significantly enhanced. Employment locally fell by only 19 per cent over the period 1980–4, compared with a drop of 33 per cent nationally. Moreover, while production employment fell by 37 per cent, salaried and managerial staff actually increased by 14 per cent. Bristol increased its share of managerial staff from 47 per cent in 1980 to 66 per cent by 1984. The decision to centralize on Bristol was influenced by the availability of suitable premises at the new production site, by Bristol's acceptability to existing management and administrative staff compared with Nottingham as the main alternative, and by the fact that Bristol was the 'natural centre' for the group as a whole. More recently, the takeover by Hanson's has raised the possibility of some job losses locally. In Food, drink, and tobacco, more generally, Cadbury–Schweppes' survival plan combined investment of £34 million with 700 redundancies, while Robertson's Foods responded to falling demand by closing a local jam factory employing another 700 to concentrate production on Manchester.

Paper and packaging Paper and packaging originated with, in particular, the local tobacco industry and, until very recently, maintained corporate ties with it. It prospered more generally with the post-war boom in consumer goods and developments in packaging and marketing. Subsequent decline reflected problems afflicting the industry as a whole: these included competition from integrated pulp and paper producers in Scandinavia who capitalized on cheap energy, local raw material supplies, and the end of tariff barriers which protected European markets, and declining consumer demand for paper and packaging products in general in the recession coupled with longer-term decline in demand for established product lines. Demand declined sharply in the late 1970s, exacerbated by high domestic exchange rates cheapening bulk imports. Employment locally collapsed from 11 500 to 5000 over the decade to 1981, with further contraction since. This was reflected in the closure of a board mill with the loss of 1700 jobs and major contraction of surviving work-forces. Less profitable product lines were eliminated, new investment to increase productivity on continuing production displaced additional employment, and limited diversification into new product lines did little to offset job loss. The fact that Bristol employment included the headquarters of a major multinational gave some stability to employment, although this too was affected by job shedding. Decline in paper and packaging had knock-on effects seen in the closure of a factory employing 600 people making machinery for the paper and board industry.

Aerospace Compared with the other main manufacturing sectors, decline in aerospace employment through the 1970s was modest and, indeed, recovered towards the end of the decade. Accounting directly for nearly 22 000 in 1981, it played a leading role in shoring up manufacturing employment as a whole, more than a quarter of which was in aerospace. Employment trends have reflected, in particular, the fortunes of the three major employers: Rolls-Royce Aero Engines, and the Aircraft and Dynamics Groups of British Aerospace (BAe) operating in separate but related markets. Thus, job loss in the early 1970s largely reflected the run-down of employment in the BAe Aircraft Group following the demise of Concorde. Within the Group as a whole, however, new prospects were found and work was allocated to the Bristol site; though half its peak in the early 1970s, employment is now relatively secure. Reinforcing this, and consolidating the site's corporate status, project management for the Airbus project was established at Bristol in 1984. Further job loss in the early 1980s largely reflected the 'recovery programme' instituted by Rolls-Royce which shed around 4000 jobs locally from 1979 to 1984. This reflected both adjustment to falling market demand—orders for civil airliners held up through the 1970s but subsequently slumped—and efforts by the UK aerospace industry generally to raise productivity levels by a combination of investment and a squeeze on manpower. Offsetting these losses, employment in BAe Dynamics, which produces guided weapons, electronic systems, and related products, increased continuously apart from a pause in the early 1980s. Consequently, employment in the sector as a whole actually increased by around 1300 jobs between 1978 and 1981.

Locationally, the sector originated in the Bristol area in the early years of aviation and was consolidated by rearmament, which included the build-up of production beyond the supposed range of enemy air attack, and by post-war reconstruction which built on established expertise and capacity. The history of corporate restructuring, locally and nationally, is complex, including public ownership of both major companies. Subsequent corporate decisions on project allocation and corporate structure, however, maintained Bristol's position in what remained essentially an expanding industry. This built on existing capacity and, in particular, the technical and scientific skills of the established work-force. Significant factors have been: the establishment of guided weapons development which formed the basis for the growth of BAe Dynamics; the allocation to Bristol of the bulk of UK Concorde work—the airframe to BAe and engine to Roll-Royce (then Bristol Siddeley)—which, though foundering in the 1970s, sustained the sector before that; and the market success of specific, locally based products, including guided weapons systems such as Sea Wolf and the RB 199 and Pegasus engines for the Tornado and Harrier. Stability of markets and employment has been increased both by corporate strategies of spreading work across sites and by the fact that development and production for both military and civil products are now largely undertaken through international collaborative projects underpinned by state support in order to spread costs, standardize defence systems across NATO partners, and extend production runs by accessing a range of national markets (Lovering 1985, 1986).

Defence markets and the particular relationship implied with the government have been fundamental to the sector. Defence equipment nationally accounted for over 70 per cent of BAe sales and 60 per cent of Rolls-Royce's in 1984; UK government contracts accounted for 63 per cent of this in the case of BAe and over 50 per cent for Rolls-Royce, the balance in each case being for export. This has been reflected locally in a succession of projects across all three companies, with BAe Dynamics, in particular, engaged primarily in defence work. Available information suggests that, while the largest share of procurement in absolute terms goes to the South East, the South West benefits disproportionately in comparison with its contribution to overall manufacturing output. Over the period from 1974–5 to 1977–8 for example, the South West accounted for only 5.5 per cent of manufacturing output, but over 10 per cent of Ministry of Defence procurement by value (Table 2.6). Defence contracts were equivalent to 10 per cent of manufacturing output compared with 8 per cent in the South East and just over 1 per cent in Wales. This excludes export sales, likely to be even more biased to major production sites such as Bristol. More recent figures suggest that this pattern continued into the 1980s. In 1983–4, defence equipment expenditure was equivalent to over 4 per cent of regional manufacturing output in the South East and South West. This contrasts with under 1 per cent in Wales and Yorkshire and Humberside, and less than 2 per cent in th North, Scotland, the West Midlands, and Northern Ireland (Breheny 1986). Domestic defence procurement and export sales have thus been a major and sustained source of demand. Moreover, projects typically extend over several years and a succession of developments, and bring related support, maintenance, and spares work. Research and development (R&D) costs are largely financed by the government, and exports are largely of products developed under contract to the Ministry of Defence and are heavily supported by government in financial and marketing terms including, for example, soft loans to purchasers or reciprocal trade agreements.

Also of major importance has been government finance and support for specific civil aerospace projects—in particular Concorde and, most recently, the Airbus—via the Department of Trade and Industry. Development and production finance for Concorde, largely benefiting Bristol, totalled nearly £1300 million over the decade to 1981–2, mainly in the early 1970s.[7] This is on the top of spending on R&D and sectoral assistance which has benefited aerospace more generally, plus domestic and export sales support.

High-technology industry Electronics and 'high-tech' industry are integral to Bristol's M4 Corridor image and to optimistic parallels drawn with California's Silicon Valley. The area boasts Hewlett Packard, manufacturing computer peripherals and establishing their first R&D facility outside the United States of America (USA) on the city's northern fringe, and the R&D facility of the microchip company Inmos. Dupont are soon to start producing electronic components for the UK market locally. Several smaller companies are

Table 2.6. *Defence Procurement and Regional Assistance by Region, 1974–1975 to 1977–1978*[a]

Standard region	Net manufacturing output (£m.)	Regional assistance[b] (£m.)	Defence procurement (£m.)	$\frac{\text{Procurement}}{\text{Output}} \times 100$ (%)	Regional assistance + procurement (£m.)	$\frac{\text{Assistance + Procurement}}{\text{Output}} \times 100$ (%)
North	10 478	476	573	5.5	1 049	10.0
Yorkshire and Humberside	15 393	89	245	1.6	334	2.2
East Midlands	12 047	10[c]	744	6.2	754	6.3
East Anglia	4 673		264	5.7	264	5.7
South East	46 726		3 674	7.9	3 674	7.9
South West	*9 078*	*26*	*889*	*9.8*	*915*	*10.1*
West Midlands	20 531	2	599	2.9	601	2.9
North West	23 920	250	949	4.0	1 199	5.0
Wales	7 448	239	112	1.5	351	4.7
Scotland	14 225	424	549	3.9	973	6.8
Great Britain	164 519	12 516	8 598	5.2	10 114	6.2

[a] Sum of totals for financial years 1974–5 to 1977–8 in current terms, except net manufacturing, calendar years 1974–7.
[b] Regional Development Grants plus selective regional assistance.
[c] Includes assistance not split between East and West Midlands.

Source: Regional Studies, CSO, net manufacturing output and regional assistance; Short (1981), defence procurement.

manufacturing electronic capital equipment, and a range of computer-related companies have taken premises on the Aztec West business park at the M4–M5 interchange to the north of the city, including Digital, ICL, GEAC, and Systime.

One definition of high-technology industry is in product terms on the basis of official Census of Employment categories.[8] Excluding for the moment aerospace, in 1981 high-technology employment—mainly electronic components, computers, and capital goods—actually accounted for a lower proportion of jobs in Bristol than nationally—only 0.8 per cent of total employment (2500) compared with 2.0 per cent. Employment locally did more than double over the decade to 1981 and survey evidence suggests that subsequent establishment of new firms and the expansion of existing companies may have added a further 800–1000 jobs since then. Whilst the sector as a whole has expanded rapidly, growth in absolute terms remains modest, particularly set against manufacturing job losses in other sectors, or service sector growth. Several companies on the Aztec West development are, moreover, essentially regional sales and service centres, capitalizing on Bristol's location and communications, but not unique to this area. Job creation in individual enterprises has also generally fallen far short of enthusiastically inflated claims. An avionics company which early press reports suggested would bring 1000–1400 jobs employed only 220 by mid-1984, with no significant plans for expansion. Hewlett Packard, expecting to employ 1300 by the mid-1980s and up to 6000 by the mid-1990s when its site was fully developed, actually employed around 300 by mid-1985 and had increased this to 510 by the end of 1986. Set against this, Sperry Gyroscope, an electronics-related 'catch' for the city in the 1970s, in fact closed down its Bristol operation in 1981 with the loss of 230 jobs locally. Moreover, the area has not attracted the relatively large-scale employment of the mass-manufacture electronics companies such as Japanese and American plants set up, typically, in the Assisted Areas of South Wales and Central Scotland

Technologically advanced activity and employment is in fact dominated locally by aerospace—'Hypertechnology' as BAe publicity has described it—rather than by more recently established companies such as Hewlett Packard, whether in terms of products, process, or employment structure. Activities within BAe Dynamics, in particular, include large-scale R&D and production related to electronics-based systems and sub-assemblies for guided weapons, space and communications, and related product areas. The plant includes a bulk printed-circuit-board manufacturing facility, and a research centre opened in 1983 is intended as the lead site for R&D work in the group as a whole. Employees classed as scientists and technologists accounted for 30 per cent of total employment and technicians for half as many again, and locally the sector as a whole employs at least 3000 people qualified to graduate level or above in electronics-related areas—more than total high-technology employment outside of aerospace. The Dynamics

Group in particular has been well placed to take advantage of the dramatic growth in government spending specifically on defence electronics which, excluding R&D, increased its share of total defence equipment expenditure from 25 per cent in 1978–9 to an estimated 33 per cent by 1983–4 (Dodd *et al.* 1983). The other main aerospace companies are similarly involved in technologically advanced R&D and production, though primarily engineering and materials based rather than electronics. Aerospace companies have, moreover, invested heavily in advanced process technology including computer-aided-design and manufacture, 'pick-and-place' robotics, and Flexible Manufacturing Systems. This contrasts with much of non-aerospace high technology which largely involves more basic assembly and wiring tasks.

Aerospace has, moreover, influenced other high-technology and electronics activity locally through purchasing and subcontracting and—more importantly perhaps—its structuring of the local labour market. A number of companies are oriented, in part at least, to local aerospace markets or have developed on this basis. Others have drawn on the pool and tradition of production labour with electrical and electronics skills as well as technical and managerial expertise built up in aerospace. This, together with the general tradition of skilled manufacturing work, has been seen as a specific positive attraction by incoming firms such as Hewlett Packard—an attraction emphasized by the key role of technical and scientific labour in determining competitive strength in the electronics industry. Reflecting this, recruitment is frequently targeted on the locality by firms further down the M4 Corridor, elsewhere in the country, and even the USA.

Aerospace and electronics have not, however, been the focus for the sort of symbiotic generative growth process around emerging dominant new technologies of the type frequently ascribed to California's Silicon Valley (Saxenian 1980, 1983; Markusen 1983). Purchasing and subcontracting by aerospace, though important locally, is relatively small scale and unidirectional, much of it going to the South East and West Midlands; there are instances of new companies spawned by aerospace or drawing key personnel from the sector, but again this is relatively small scale. In particular, the infrastructure of electronic capital equipment manufacture characteristic of Silicon Valley and associated with a particular phase of technological development is relatively undeveloped in the Bristol locality and the M4 Corridor in general. Finally, the sorts of academic–industry relation implicated in the genesis of Silicon Valley have failed to materialize, reflecting differences in corporate tax structures, the significant role of government research establishments, and the internalization of other research activity by the aerospace industry itself. Over the two years 1982 and 1983 for example, only a couple of minor projects were funded by private sector aerospace and electronics companies at Bristol University, and support from the Ministry of Defence totalled only £300 000 per year.[9] Parallels with Silicon Valley in general, then, owe more to marketing hype than high-technology expansion itself.

Financial services Employment in financial services, totalling over 22 000 by 1981, partly reflects Bristol's role as a major urban and regional centre. A key component in the employment growth of nearly 9500 over the decade to 1981 was, however, relocation, primarily from London. Four UK insurance company administrative headquarters, a finance company, and several departments of one of the 'Big Four' banks have moved to Bristol since the early 1970s, including most recently London Life Assurance. This represents essentially 'basic' employment, 'exporting' services to the rest of the country and generating, rather than being dependent upon, local economic activity. Insurance activity, moreover, is primarily in the life and pensions sector which has remained relatively buoyant despite the recession. Companies relocating were mainly seeking to reduce the cost of office space, particularly when faced with the need to expand and consolidate existing office space. Office rents in Bristol in the early 1970s were under 10 per cent of City levels and 15–20 per cent of West End levels, although the difference narrowed a bit later.[10] High turnover, low quality, and shortages of staff in London were, in the early 1970s, a subsidiary factor. Bristol combined availability of low-cost sites and premises; a large enough employment pool beyond the London catchment area including high-quality school-leavers, female workers, and, as the sector built up, experience of insurance work; and access, primarily to London, by motorway and High Speed Train. However, particularly important in narrowing the choice down to Bristol was its attractiveness to sufficient numbers of key staff to make moves viable—a combination of factors including quality of life, educational, recreational, and cultural facilities, housing and the proximity, both physical and psychological, to London and the Home Counties. Property consultants Knight, Frank, and Rutley observed that Bristol was

> probably the only serious relocation candidate which combines the advantages of a major urban and office centre, with a location in a very attractive part of the country. In addition, it has ample supply of new offices in different sizes and locations. The relatively long distance from London is compensated for by superior communications and low accommodation costs relative to centres nearer London. (Knight, Frank, and Rutley 1983.)

None of these factors was sufficient in itself; nor did they uniquely identify Bristol. Other financial institutions moved to other localities; but the concentration of insurance company relocation was remarkable. Relocation and the scale of the financial services sector, moreover, had the cumulative effect of stimulating a range of activities, such as insurance broking, ancillary to the main employers.

By 1983, employment in the five main relocated companies was over 4000. About half the labour force initially relocated with their company, tending to be older, male, and in higher clerical, technical, and managerial positions. In one case, only 9 per cent of those relocating were under twenty-five compared with 60 per cent of locally recruited staff, 61 per cent earned at least £10 000 compared with 10 per cent of local recruits, 33 per cent were in managerial positions compared with 3 per cent of local recruits, and 77 per cent were male compared with 44 per cent of local recruits. Local recruitment was mainly typists, lower clerical grades, and ancillary employment, representing an important source of female employment. Women represented over half of total employment in financial services in 1981, including 3500 in insurance alone, although the type of employment is significant. The majority are employed mainly in lower clerical, typing, and secretarial grades. In one case, women accounted for 70 per cent of the bottom three clerical grades but only 25 per cent of the top three, with virtually none in managerial and senior technical positions. Despite major contrasts in terms of gender within the sector, financial services do, nevertheless, provide relatively secure full-time female employment with good pay and conditions compared to much of the private service sector.

Miscellaneous services The other major source of service employment growth has been miscellaneous services, where over 80 per cent of the growth has been represented by growth in female employment. The situation obviously varies across different activities, but, as already suggested, much of this is part-time employment. Parts of miscellaneous services are, moreover, characterized by relatively poor pay and conditions and little protection from formal work-place organizations, verging, in some cases, on casual labour. The sector includes a wide range of mainly personal services including, for example, pubs and clubs, restaurants, cleaning and repairs, and many other services not individually identified by the official figures. These activities are essentially 'dependent' on the prosperity of other sectors in the economy. In particular, the scale and character of their growth appear to reflect demands generated by the growth of relatively well-off salaried, professional, and technical workers in, for example, financial services and technologically advanced manufacturing. They also reflect less tangible changes in patterns of demand and life-style. This has been reinforced in the case of Bristol by a distinctive interrelated growth of leisure, business, and tourist activity focused on the central area and the redevelopment of the city docks.

It is therefore evident that employment locally has been affected by a variety of processes specific to particular sectors and corporate structures, operating on national and international scales. The decline of paper and packaging locally, for example, reflects specific industry-wide pressures; the tobacco sector has been affected both by falling demand in general and the specific corporate circumstances of the industry locally; aerospace employment has reflected continuing corporate commitment to existing capacity, labour, and expertise, together with overall market strength and the success of specific products; and insurance decentralization has reflected the relative cost of office space in central London. The area's specific legacy of previous rounds of investment in terms of current economic and employment structure is obviously central to any understanding of the contemporary situation. A number of more general factors can, however, be drawn out.

First, access to London, Heathrow, and the rest of the country—including the early introduction of the High Speed Train—has been important, particularly to the financial services and electronics sector. This has not only facilitated business contact and movement of goods, but also functioned to tie in the locality to London and the rest of the South-East 'heartland', easing the transfer in, and attraction, of higher-grade staff. Regional nodality, also, has reinforced Bristol's role across a range of distribution and service functions. Second, the scale and diversity of the local labour market, including higher-grade technical and professsional staff, skilled manual labour, female employment, and good-quality school-leavers, has been important. This has been reinforced from the employer's perspective by a good industrial relations reputation. Third, the scale of the urban environment and rural surroundings, housing, education, leisure, and cultural facilities—the consumption, life-style, and image factors—have facilitated the recruitment and retention of higher-level technical, professional, and managerial staff. These factors have also been influential more generally in terms of locational and investment decisions. In contrast to other major conurbations, Bristol has inherited little in terms of physical dereliction or image from its industrial past, with the city centre docks, virtually deserted by commercial traffic by the 1960s, providing instead the opportunity for what has been seen as showpiece renovation and redevelopment for arts, leisure, and recreational uses and the revival of city centre private housing development. Less tangible, but important nevertheless, the locality has a positive social image for elite labour groups and corporate decision-makers, and lacks any element of stigma attached to designated inner city and regional 'problem areas' (Massey and Miles 1983). Finally, the cost and availability of land and premises, including both central area office space and major green-field sites, particularly on the northern fringe, have facilitated the growth of office-based services, including relocation of financial services, and the establishment of financial services, and the establishment of electronics-related activities. Speculative over-provision in the early 1970s, unhindered by planning constraints and, if anything, intensified since, has led to the city's locational attractions being 'undervalued' in terms of office rents (Knight, Frank, and Rutley 1983). The amenability of the local planning environment and the local authority environment was a factor in some instances. While many councils would welcome the opportunity to respond to the sort of growth options represented locally, Knight, Frank, and Rutley have, for example, referred specifically to Bristol's 'positive office planning policy'. Moreover, in the M4 Corridor context, Bristol and Northavon Districts, and Avon

County, have in general terms been more pro-development than, in particular, much of Berkshire.

Employment prospects In terms of future employment change, county council estimates suggest that over the decade 1981–91 employment as a whole will increase by around 4 per cent, with a manufacturing decline of around 5 per cent (5000) being more than compensated by an 8 per cent (21 000) increase in services.[11] In much of manufacturing, including tobacco, and paper, printing, and packaging, the major impacts of the recession, restructuring, and job shedding have largely worked their way through. Employment is now more stable, and only gradual shrinkage is likely, in the medium term at least. Aerospace employment, in particular, is firmer now following losses since 1981, and may even show some increase, in line with national estimates. In the longer term, employment is dependent on the success of individual projects such as the Airbus and, in particular, on the level and structure of defence expenditure. Defence equipment expenditure may well stabilize in real terms. Industry analysts have, nevertheless, concluded that 'The electronics suppliers will, however, continue to benefit from the swing away from people and ironmongery to 'clever equipment' using advanced electronics' and that 'In a nutshell, on the basis of both global and UK considerations, the defence electronics sector looks gold-plated' (Dodd *et al.* 1983). Locally, BAe Dynamics, in particular, is likely to benefit from this trend. Some further growth in other high-technology employment is likely, both in existing employers and through the attraction of new firms. However, this is unlikely to be very dramatic in absolute terms. Only modest growth and inward investment are forecast nationally, and the area faces competition from the whole Cambridge–London–M4 belt. Growth in aerospace and other high-technology employment may altogether add around 2000 jobs over the decade to 1991. Growth of financial services is expected to continue, but less dramatically than over the previous decade, adding around 6500 jobs locally by 1991. Further relocation remains a real possibility. Bristol remains attractive should firms be seeking to relocate. Dispersal rates are obviously well down on the early 1970s, but Jones, Lang, and Wooton (1983), for example, suggested in June 1983 that twelve office moves involving over 100 employees were likely in the immediate future, with another twenty in prospect. Major growth of miscellaneous services is also forecast, adding around 10 000 jobs over the decade to 1991, particurlarly in leisure-related personal services dependent on the continued prosperity of basic employment, but also related to the growth of tourism and business-related activity. Little employment increase is likely elsewhere in the service sector, however, with some losses in public utilities, distribution, and possibly transport and communication. Public service employment as a whole is likely to decline, given continuation of current public expenditure restraint, with modest growth in education and health having tailed off. Construction employment is similarly likely to remain depressed unless public investment in infrastructure and housing is increased.

Corporate structure Looking, finally, at some limited evidence relating to employment change in terms of ownership and corporate structure, non-UK ownership does not appear to be particularly implicated in employment decline. Employment in foreign-owned manufacturing companies locally declined by 12 per cent compared with 9 per cent for UK companies, from 1975 to 1978 but, as the recession deepened, the position was reversed, with employment in UK-owned companies falling by 13 per cent and that in foreign-owned companies by only 3 per cent. In terms of indigenous versus externally generated growth, over the period 1972–83, manufacturing firms moving into the locality employed 1320, new branches established locally employed 1600, and new single-plant enterprises employed 750, suggesting that locally generated manufacturing employment was of relatively little importance. Moreover, only new branch plants showed a net increase in employment between the time they were established and the end of the period.[12] Roughly speaking, the scale of 'external' ownership[13] in Bristol is not particularly distinctive compared with conurbations in the Midlands and North (Bassett 1984). In terms of firm size, smaller manufacturing establishments, employing 11–50, showed a significant growth rate, contrasting with job loss in all other size bands and distinguishing Bristol from Clydeside, the West Midlands, and London. However, growth was too modest in absolute terms to suggest that small firms were particularly significant.

To summarize, and draw out a few more general conclusions, manufacturing job loss has hit certain sectors hard since the early 1970s. However, decline generally came later and was less severe than in many localities which have suffered long-term decline in basic industries, or compared with the more recent collapse of engineering and metals-based industries as in the West Midlands. Aerospace, in particular, has shored up employment, maintaining a dynamic core of technologically advanced activity and employment with wider economic and employment implications. Decline in other sectors has represented delayed restructuring and 'shake-out' of employment rather than collapse and has left a core of more stable higher-productivity activity in tobacco, printing and packaging, and other industries. While forced redundancies have occurred, a significant feature of employment decline locally has been the extent to which it has been achieved by a combination of 'natural wastage' through non-replacement as vacancies arise, early retirement, and 'voluntary' redundancy.

As elsewhere, manufacturing job loss has been offset by a major expansion in service employment, but with financial and miscellaneous services in particular growing much faster than nationally. Strong growth in public services reinforced the continuing strength of much of manufacturing through the

1960s, to be overtaken in the 1970s by private service expansion. Accelerating job loss and unemployment in the mid-1970s relative to the national picture reflected the scale of manufacturing job loss in the three main sectors, including in the early 1970s aerospace, temporarily breaking through' the offsetting effects of service expansion. Growth in financial services largely reflected relocation of insurance activity from London, while miscellaneous services have responded to both quantitative and qualititative changes in local demand. This has been emphasized by the maintenance and expansion of relatively high-income professional, technical, and other white-collar strata in the labour force in aerospace, electronics, financial services, and related activities. For the future, manufacturing employment as a whole is likely to be more stable, but with aerospace remaining of major importance. Significant employment growth is likely to be confined to financial and miscellaneous services, with public service employment likely to decline under the current public expenditure scenario. High technology is unlikely to be a significant source of new employment in absolute terms.

Finally, in terms of corporate role and function Bristol's position was consolidated and extended. Insurance relocation established major headquarters functions in the city. Centralization of financial services, and other service activity more generally, emphasized Bristol's wider 'regional' function extending in some cases to South Wales, the South Coast, and much of the M4 Corridor in some cases. Headquarters functions in tobacco were centralized on Bristol. Corporate restructuring and project allocation, including Airbus management, BAe Dynamics' R&D role, and Rolls-Royce's role in both R&D and the Military Engine Group, if anything, strengthened the place of local aerospace activity in the wider company context. Lastly, Hewlett Packard's R&D facility has been added to production capacity, reinforcing the modest but potentially important presence of non-aerospace R&D locally. At a more general level, there is little to suggest that the locality's economic fortunes reflect a distinctive structure of corporate control in terms of the importance of, for example, branch plants, external or non-UK ownership, or firm size. The Bristol case suggests that external control—sometimes argued as a key factor in regional decline—is simply a channel through which a variety of corporate strategies may impact, positively as well as negatively in any particular case. It does not, moreover, support the argument that small firms have a particularly important role in employment generation.

Employment and Unemployment

Bristol's work-force continued to expand through the 1970s, albeit more slowly than in the boom years of the post-war period when in-migration to the county as a whole often exceeded 3000 per annum,[14] Fewer workers—under 1000 a year—were moving in during the 1970s, while the overall activity rate, which had been rising, changed little. The number of school-leavers entering the work-force has, however, been well up on the number of people retiring since the late 1970s. This left the resident work-force in 1981 around 18 000 up over the decade—an increase of 5 per cent,[15] contrasting with decline in the conurbations and most larger cities elsewhere. With the female activity rate still rising and the male rate falling, however, the proportion of women in the work-force increased more rapidly. Over the longer term, Bristol was one of only a handful of the largest urban areas in which employment was greater in 1981 than in 1951.[16]

There was an element of core decline and peripheral expansion, with Bristol District losing around 4 per cent of its resident work-force and the rest of the area gaining nearly 8 per cent over the decade to 1981.[17] This relatively minor shift largely reflected the growth of commuting patterns rather than, in any sense, central area economic decline relative to the fringe. New housing growth has been concentrated in the conurbation fringe, in 'overspill' estates, and in growing 'dormitory' towns and villages—a pattern which the projected growth in the county's resident work-force by over 13 000, between 1981 and 1986 will reinforce.[18] Incoming workers in, for example, financial services have, in particular, supported dormitory growth. However, commuting journeys are short compared to those in much of the South East, reducing the social and economic dislocation between residential and urban areas.

Overall expansion in the work-force has been reflected in unemployment totals in the locality, although the relationship between employment, unemployment, and the overall work-force is complex. In particular, increased employment tends to increase activity rates, especially for women, and is not therefore directly reflected in unemployment figures. Unemployment in the locality, at 10.8 per cent in January 1987, compared favourably with the national figure of 13.2 per cent. Throughout the long boom, the area had experienced effectively full employment, with job growth keeping employment below 10 000 in the county as a whole and the rate fluctuating between 1 and 3 per cent. Unemployment locally then climbed sharply in the early 1970s in line, initially, with the national picture (Figure 2.3). In late 1975, however, it overtook the national rate and remained marginally higher until early 1980, reflecting the localized 'economic crisis' referred to earlier, when manufacturing job loss moved sharply upwards. As the recession deepened, however, and unemployment nationally accelerated, Bristol's position relative to the rest of the country improved significantly. The rate of increase in unemployment slowed considerably over 1984, although the total continued to expand. In the 'M4 Corridor' context, however, unemployment is relatively high compared with, for example, that in Reading, although it is comparable with that in Swindon and well down on the rate for Cardiff. And, while the area has in no sense escaped the impacts of recession, the scale of unemployment remains low compared to, for example, the scale in Glasgow,

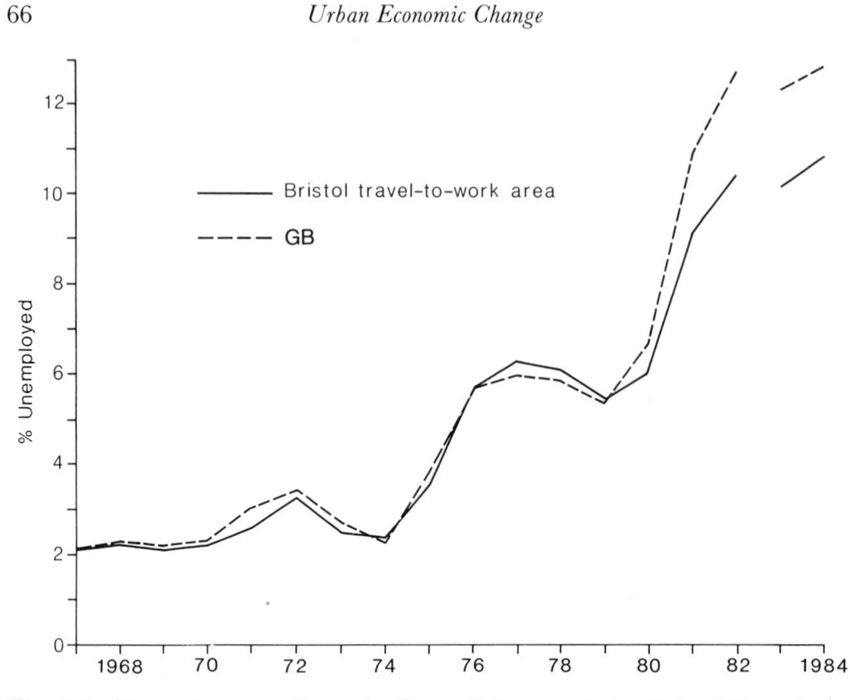

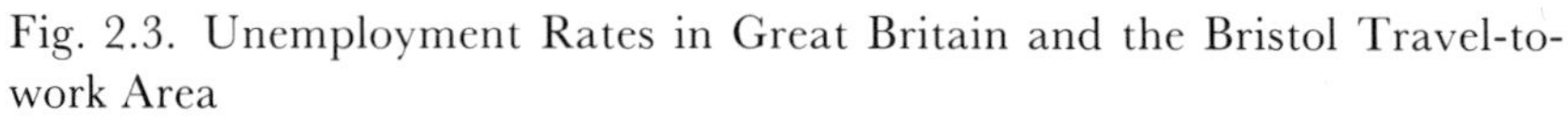

Fig. 2.3. Unemployment Rates in Great Britain and the Bristol Travel-to-work Area

Source: Monthly unemployment figures, the MSC, and the Department of the Environment.

Table 2.7. *Unemployment in Selected Localities, 1978 and 1986*

Locality[a]	Oct. 1986 (%)	Jan. 1978 (%)	% increase in rate
Reading	7.1	4.2	69
Swindon	11.2	6.9	62
Bristol	10.7	6.8	57
Cardiff	14.0	7.7	82
Glasgow	17.2	10.3	67
Newcastle	17.9	9.4	90
Birmingham	16.1	6.3	156
Great Britain	12.9	5.9	119

[a] Travel-to-work areas.

Source: Department of Employment Gazette.

Newcastle, and the West Midlands where unemployment increased sharply from the late 1970s (Table 2.7).

The persistence of unemployment for particular groups in the labour market, and the scale of unemployment among ethnic minority groups and young people is, however, particularly marked.[19] The proportion of the unemployed who had been out of work in the county for over a year increased from 14 per cent in January 1976 to 36 per cent (16 700) by October 1984, little short of the national figure of 40 per cent. Long-term unemployment among older workers was in fact significantly higher locally than nationally, with 59 per cent of those out of work in the 55–59 age group having been so for over a year compared with 52 per cent nationally. Unemployment flow figures, however, indicate that, on average, those looking for work in Bristol are unemployed for a slightly shorter period than nationally, particularly younger workers.[20] Many in fact remain unemployed for relatively short periods of time, suggesting an increasing polarization between the long-term and more transitory unemployed.

Unemployment among minority ethnic groups, who make up 2.5 per cent of the total work-force, was just over 1 percentage point up on the overall rate in August 1982.[21] However, the unemployment rate for male workers of West Indian origin was 24.0 per cent, compared with 12.4 per cent for male workers as a whole.[22] Male workers of West Indian origin accounted for only 1.1 per cent of the male work-force, but for 2.2 per cent of male unemployment.

Youth unemployment has risen consistently as the economy locally has failed to absorb the influx to the labour market of school-leavers. The unemployment rate for those under 25 years old in the county had reached nearly 19 per cent by October 1984, compared with 11 per cent overall. This was despite the impact of the Youth Training Scheme (YTS) and the Community Programme, without which the rate would have increased to around 25 per cent. Moreover, 29 per cent of 20–24 year-olds had been unemployed for at least a year. Youth unemployment rates locally equalled national rates, except for the youngest, 16–17 year-old, group. Indeed, young people in Bristol bear a marginally larger share of the burden of unemployment than elsewhere, under-25-year-olds accounting for 34 per cent of the unemployed locally compared with 31 per cent nationally. While older workers are less likely to be unemployed, they are likely to remain out of work for longer, 55 per cent of the unemployed aged 45–59 having been so for over a year. So, while unemployment overall is below the national figure, long-term and youth unemployment compare less favourably.

This is indicative of a growing mismatch not only in terms of numbers of jobs but also in terms of the types of employment offered by the expanding sectors of the economy identified earlier. Job loss has been concentrated on manual employment, much of it semi-skilled and unskilled and, typically, male employment. Employment growth, on the other hand, has been mainly in the service sector, much of it offering job opportunities to women, frequently part-time. There is a sense in which, as traditional manufacturing declined and services expanded, male manual workers, particularly older workers, moved into unemployment, frequently long term. On the hand, women, including both new entrants and returners to the labour market together with some previously employed in manufacturing, were drawn into the rapidly expanding service sector. Service sector growth has clearly not provided job opportunities for those losing jobs in manufacturing or who might otherwise have looked to the sector for employment. Moreover, in the more resilient manufacturing sectors, such as aerospace and the newer electronics companies, the qualifications, technical knowledge, and experience required are major barriers, with many jobs being filled by those already employed. In the service sector, the qualifications required for higher-grade positions within, for example, financial services similarly exclude the great majority of school-leavers and other job seekers. Unskilled and semi-skilled manual job opportunities have been at a particularly low level, in part reflecting the state of the construction industry and the constraints on public housing, environmental improvement, and infrastructure provision. It is as evident in Bristol as in less prosperous localities that the scale of construction activity, both public and private, is particularly important to the job prospects of these particular groups.

There is some evidence, also, of increasing polarization within the labour market itself. On the one hand are those in relatively well-paid and secure employment with relevant skills, technical or professional expertise, frequently some form of career structure or salary progression, and opportunities for job mobility. These would include technical, professional, and managerial staff and some skill groups in, for example, aerospace, electronics, financial services, and the public sector professions. These contrast with the much larger group including not only those in declining sectors but also less-skilled and lower-grade employees in surviving industries, and those whose employment and job prospects are tied to de-skilled and routinized work in manufacturing and, particularly, relatively insecure, low-paid, and often part-time work in the service sector. There is some evidence as well that, in absolute terms, growth has been concentrated in employment categories where average earnings are well below the national level, in both male and female non-manual employment. Female manual employment, on the other hand, where earnings are slightly above average, has been contracting, reflecting the decline in traditional areas of female employment in, for example, tobacco and packaging. Polarization is also seen to some extent within particular employment sectors. In insurance, for example, there is a growing distinction between clerical staff involved in basic data input and other routine tasks, and 'career' staff who are increasingly specializing in particular activities and drawing on new technology and who progress through the lower clerical grades. In parts of aerospace and electronics, the increasing technical and professional content of much of the work contrasts with routine assembly tasks.

On current employment forecasts, the level and structure of unemployment locally is not expected to change markedly. Despite the fall in the number of school-leavers, the numbers entering the work-force still exceed the numbers retiring, and overall numbers unemployed are likely to continue to rise. Unemployment among school-leavers may fall as their volume continues to fall and YTS and other government schemes absorb a significant proportion, extending to 18–19 year-olds subsequently. Employment prospects for the current young unemployed are unlikely to improve, and a sharp increase in long-term unemployment for those in their twenties is now becoming evident. Long-term unemployment as a whole will increase, possibly reaching 20 000 locally by the end of the decade[23] (that is, well over 40 per cent of total unemployment) and affecting in particular less skilled manual workers, who are unlikely to benefit from job opportunities for those with technical skills and in the service sector.

The overall picture in terms of unemployment, therefore, compares relatively favourably with many other localities. However, this disguises the impact of long-term unemployment on particular employment categories and the scale of youth unemployment. Indicative of broader trends, increasing polarization is evident, not simply between those in and out of work, but between managerial, professional, technical, and other higher-level staff benefiting directly from the more resilient and expanding sectors of the economy, and others in both the declining sectors and the lower reaches of the expanding 'dependent' services sector.

Spatial Inequalities and the Inner City

Overall figures for unemployment disguise marked spatial inequalities, which are also mirrored in a range of other indices of economic and social deprivation. Focusing on Bristol District, the 1981 Census shows unemployment ranging from 4 per cent in the best ward to over 23 per cent in the worst. The overall spatial pattern of unemployment picks out both an 'inner city' band of high unemployment north and east of the city centre, peaking in St Pauls Ward, and also peripheral council estates on the edge of the district—a pattern which an Office of Population Censuses and Surveys (OPCS) Report (Redfern 1982) described as typical of 'council estates in Liverpool, Sunderland and Newcastle upon Tyne' (Figure 2.4). While unemployment rates, locally, peak in the inner zone centred on St Pauls, the peripheral estates account for a higher proportion of the total number unemployed. This general pattern is reinforced in terms of other indicators of economic and social problems such as electricity disconnections, car ownership, or statutory supervision orders.[24] Inequalities in health have also been found to be strongly correlated with other indicators of economic and social disadvantages (Townsend 1984). Problems in terms of unemployment and other indices do not therefore, in Bristol as in many other localities, follow a simple 'inner city' pattern. The city council's own definition of the 'inner

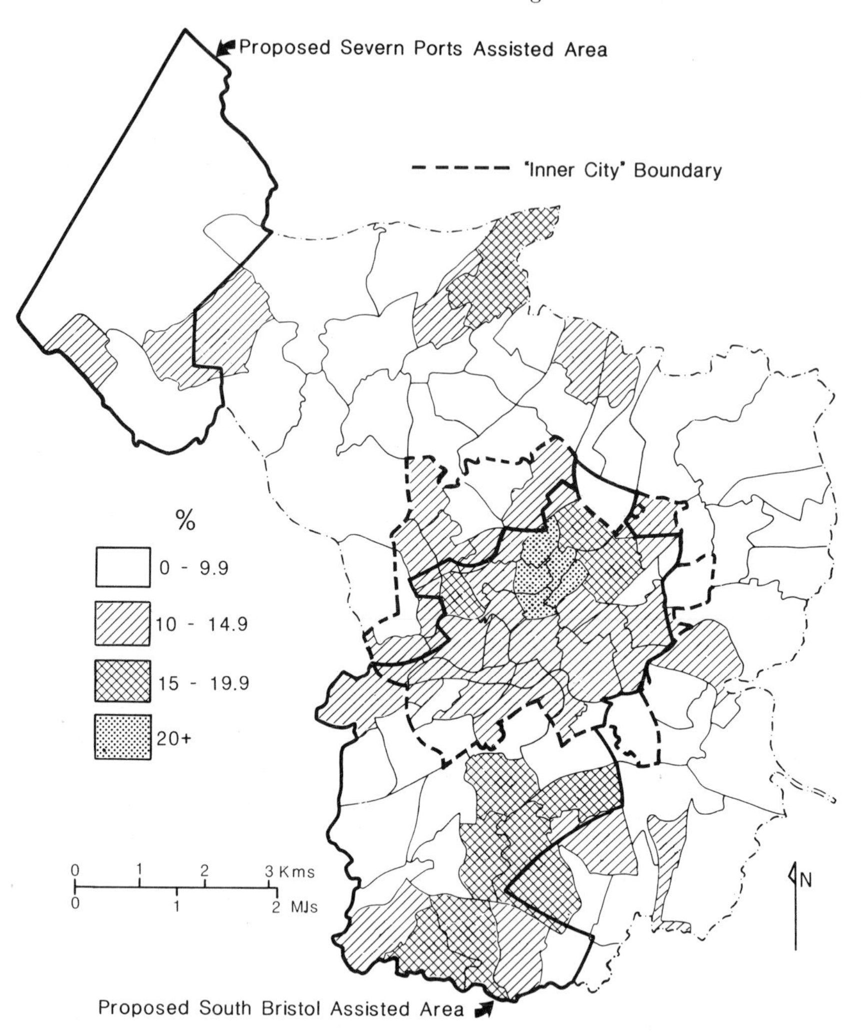

Fig. 2.4. Unemployment Rates in Bristol, 1981
Source: Department of Employment and Avon County Planning Department.

city'—an area with a population of nearly 130 000 around the city centre—in fact excludes the peripheral council estates. It also conceals marked internal diversity, including some of the most affluent areas of the city and extensive professional middle-class residential areas. Change over the period 1976–81 suggests that unemployment and other adverse indicators have tended, if anything, to intensify outside of the worst areas in 1976. There was a fairly rapid increase in unemployment on the outlying council estates, particularly in south Bristol, both in absolute terms and relative to the central area. The spatial pattern of areas with the highest unemployment has not changed markedly, however.[25]

The pattern of unemployment, adverse socio-economic indicators, and physical decay in the housing stock is thus complex. It includes a small number of inner wards close to the city centre, combining high unemployment, mixed tenure including a high proportion of rented accommodation, housing disrepair, significant black and Asian populations, and other indices typical of what is commonly understood as the 'inner city problem'. This area is not physically very extensive, however; nor does it account for a large proportion of the total population living in areas suffering high unemployment and other problems. These include at least three identifiable clusters of predominantly public housing to the north, the south, and at Avonmouth. Adverse socio-economic indicators and environmental problems here coincide with, in places, major problems of housing standards and disrepair that are seriously exacerbated by lack of local authority resources.

Spatial patterning of unemployment and related deprivation indices seems to reflect largely non-spatial processes. People are not generally unemployed because of where they live. There is evidence that employers discriminate between applicants for some categories of employment on the basis of where they live, but this is not widespread. It occurs mainly in the case of unskilled low-paid employment such as retailing where there is a large excess of applicants, and where employers may favour those from locations more readily accessible to the work-place. Problems of mismatch and accessibility have been suggested locally, between employment growth on the city's northern fringe and rising unemployment in east and south Bristol affected in particular by manufacturing job loss. However, the problem appears to reflect the type and scale of employment created, rather than, in an urban area of the scale and density of Bristol, access and the location of new job opportunities. The unemployed and poorly qualified school-leavers generally lack the qualifications and experience to take up what few opportunities arise, as evidenced by unemployment rates on council estates in the north, within sight of Aztec West and the aerospace complex, that are as high as those in south Bristol. The spatial structure of unemployment and multiple deprivation seems therefore to reflect, above all, processes of housing allocation, both public and private, linked to broader processes of social and spatial differentiation. The spatial patterning of unemployment largely reflects where those with limited power in the labour market and who are most likely to be unemployed can command housing resources.

Overall indices of unemployment and disadvantage are not particularly unfavourable, when Bristol as a whole is compared with the conurbations and many other large cities. In absolute terms, however, unemployment and adverse socio-economic conditions appear much more serious. An analysis of socio-economic conditions based on 1981 Census data (Redfern 1982) found that 62 000 people in Bristol lived in wards which exhibited 'the most adverse conditions': this is significantly more than in Coventry and Sheffield, both of which have Programme status under the 1980 Inner Urban Areas Act, and just short of Newcastle which has Partnership status. Bristol was also the only authority—apart from Nottingham (which is a Programme authority)—to have a ward displaying what this study defined as the worst combination of conditions, and yet lack Partnership status. Total unemployment in Bristol, in absolute rather than percentage terms, was roughly the same as that in Sheffield and Coventry. It can be argued, then, that, in absolute terms at least, Bristol's relative prosperity conceals unemployment and other social, economic, and physical problems comparable to those in cities with officially acknowledged 'inner city' problems. This is compounded by the spatial structure of inner-city-type problems in the city, which includes outlying pockets of high unemployment and related adverse indices, as well as a more compact inner zone with particularly severe problems, contrasting sharply with more affluent areas.

Policy Issues

In terms of explicit national-level urban and regional policy, the Bristol subregion is very much an example of 'hands-off' development. Nor has it seen the proliferation of economic development and employment agencies typical of the regions and inner cities. On the other hand, other central government policies and expenditure programmes, not specifically directed to urban and regional policy ends, have major impacts on economic development locally. As elsewhere, specific local authority economic development policies have also been established.

Urban and regional policy

For central government, the Bristol area has not been defined as 'a problem'. It is one of only a handful of major urban areas that has neither had any form of Assisted Area status in terms of regional policy, nor, until 1986, explicit 'Inner City' status under the Inner Urban Areas Act 1978. Government funding for port development was refused in the 1960s, and later bids for an Enterprise Zone and for Freeport status were turned down. Lacking Assisted Area status, the area has also been ineligible for European Regional Fund support. A certain amount of grant aid has been received for individual projects under the 'traditional urban programme' and, from 1984–5, there was modest funding for training initiatives from the European Social Fund. Overall, however, financial resources received locally have been minimal. Nor has the area shared the additional powers granted to inner city authorities or the specific development agency structures established in Scotland and Wales. The locality stands in stark contrast, for example, to parts of South Wales, less than an hour west down the M4, with the complete catalogue of Special

Development Area status, European Regional Fund money, European Coal and Steel Community funds, the Welsh Development Agency, an Enterprise Zone, the Land Authority for Wales, and, finally, a Freeport.

Comparative disadvantage in this sense has been reinforced in the past by the use of Industrial Development Certificates (IDCs) to divert manufacturing industry to the Assisted Regions. IDCs were fairly readily available in the mid-1950s but became more restricted after 1958 (Bristol City Council 1967). In 1976, for example, an IDC was refused to Toyota who wanted to build a car import and distribution terminal adjacent to the new Royal Portbury Dock, with the aim of forcing them to set up in Liverpool instead. Permission was subsequently granted for a smaller development, but this was not taken up. Government pressure in 1979 forced Inmos, the state-backed microchip company, to locate its UK production capacity in Newport even though management had selected Bristol as its headquarters and R&D base and had identified a production site on the city's northern fringe. Similarly, Marconi Avionics' development was delayed by negotiations over an IDC, which was finally granted after the change in government in 1979, by which time market conditions had led to a scaling down of investment plans. The requirement for IDCs also largely prevented the construction of speculative industrial premises. Firms wanting immediate occupation of premises were more likely to find them in the Assisted areas where such restrictions did not apply and the provision of speculative premises was, moreover, a specific strategy of development agencies and local authorities. More recently, with the shift in the economic and political climate, IDC policy has been largely relaxed. In the case of Hewlett Packard, only token suggestions of a development area location were voiced by central government which formally justified allowing development to proceed on the basis that this was 'an internationally mobile project with a specific alternative location in France' and that the company had 'firmly decided that the Bristol site is the only one in the UK which will meet their exact needs'.[26] Underlining this in fact, the government provided a £5 million grant towards the company's R&D cost.

The negative impacts of regional policy are particularly hard to pin down, but such impacts have been the cause of considerable aggravation and debate involving local councils and Members of Parliament—much of it directed at the neighbouring area of South Wales. Individual cases of IDC refusal can be identified. It is not clear, however, that IDC policy overall has had a significant negative impact on the locality. Manufacturing prosperity through the 1950s and 1960s was based more on the buoyancy of existing industrial sectors and continued investment by existing local employers, particularly aerospace. There is no evidence that significant numbers of manufacturing firms were specifically prevented from relocating into the area over this period. Employment growth through the 1960s and into the 1970s was, as we have seen, primarily service sector and office-based activity—major recent relocation has involved specifically financial services. Some companies, such as Inmos and Sperry Gyroscope, were in fact able to take speculatively built central area office premises rather than designated industrial space. Finally, the council had been advised, when developing its promotional strategy, that overseas investment would not be obstructed by IDC policy.

To the extent that the positive incentives offered by the regions have been effective, then areas such as Bristol have potentially lost out. There has been much debate over the effectiveness of regional policy, with some claims of significant job creation in the Assisted Areas and diversion of manufacturing investment, particularly in the period 1967–71 (Gudgin *et al.* 1982; Rees and Miall 1981). It is particularly difficult, however, to establish the extent to which, in the absence of such measures, companies which did set up in the regions would have specifically relocated to, or expanded employment in, the Bristol locality. Certainly Bristol did not, by and large, attract the manufacturing branch plants which were established in the regions through the 1960s, nor the Japanese and American companies which set up in Wales and Scotland. The labour market characteristics of these areas were, however, a major factor: they had cheaper labour than in Bristol, (including in particular women with little tradition of industrial working) and were high-unemployment areas, thus offering a particularly favourable context within which to establish new forms of work control and management–labour relations (Morgan and Sayer 1984). Regional policy to some extent 'succeeded' by raising female activity rates rather than by reducing unemployment. Financial incentives have commonly been referred to as 'a bonus' alongside other locational factors. Moreover, low start-up costs, including the availability of premises on advantageous terms, have encouraged marginal production units, particularly branch plants, to expand into the spatial periphery. Such employment has frequently proved to be relatively unstable, even in supposed growth sectors. As international electronics markets weakened in 1985, National Semiconductors, for example, laid off 400 workers at its Greenock plant in mid-1985 following a slump in demand for microchips, and Inmos shelved expansion plans for its South Wales production unit where employment is still far short of forecast levels.

In terms of service sector expansion, Bristol in fact benefited from central government's strategy of office dispersal initiated in the 1960s. Commercial development in London was restricted from 1965 by the use of Office Development Permits (ODPs), subsequently extended to cover the rest of the South East and the Midlands. Given its attractions discussed earlier, Bristol was particularly well placed to benefit from the displacement of both development pressure evidenced in the local office building boom, and office-based employment itself. There was no effective constraint imposed on the scale of development locally. Moreover, the effects of ODPs, reinforced by speculative over-provision in Bristol, emphasized the price differential and cost advantages offered by Bristol. The related dispersal of government offices was mainly intended to relocate major employment in, for example, the Department of Health and Social Security (Newcastle) or the Driver and Vehicle Licensing Centre (Swansea) to the regions. Bristol, however, made some gains through the establishment locally of the Department of the Environment Planning Inspectorate and the Countryside Directorate in the city, employing around 900 people.

Implicit policy impacts

Also relevant, however, are the impacts of national government programmes and expenditure which, although not directed explicitly to urban and regional policy ends, nevertheless have spatially differentiated impacts. These are many and varied, and difficult to quantify with any precision, but nevertheless of major importance. Two areas in particular appear to be relevant in the case of Bristol. First, government-funded transport infrastructure gave the area excellent communications relatively early on, particularly to London and Heathrow but also to the rest of the country. As described earlier, this has facilitated movement of goods, imports and exports, sales and service activity, and contact between key personnel. It has also been a factor in locational and investment decisions, and important to the relocation, attraction, and retention of key staff. Developments include the completion of the M4 link to London with the M4–M5 interchange on the city's northern fringe, and the build-up of High Speed Train services to London from the early 1970s. The Severn Bridge link appears to have strengthened Bristol's relative position, facilitating distribution and servicing of South Wales from the English side of the border. Bristol's regional nodality, as well as national accessibility, has been greatly enhanced, and the area adjacent to the M4–M5 has been building up regional and national distribution activity linked in part to dock facilities. By the 1980s, however, it was being argued that government restrictions on local authority infrastructure spending were inhibiting the release of significant industrial sites and the possibility of prestige developments in the high-demand northern fringe (Avon County Council 1985). This included both off-site infrastrucure and delays to the Avon Ring Road extending motorway access to east and south Bristol.

Second, as already indicated, the locality has benefited in particular from defence procurement, and from support for both defence production and civil aerospace projects. Aerospace as a whole has, as shown earlier, had major employment, labour market, and wider, multiplier, effects. A crude measure of the significance of defence procurement in terms of regional prosperity is possible by comparing it with explicit regional policy expenditure. In Wales, for example, from 1974–5 to 1977–8, regional assistance amounted to £239 million, compared with only £26 million in the South West (see Table 2.6). Ministry of Defence procurement contracts in Wales, however, amounted to only £112 million, compared with £889 million in the South West. Defence procurement spending thus far outweighed in straight financial terms any comparative advantage established for neighbouring Wales by explicit regional policy. Regional assistance and defence procurement together represent a higher proportion of regional output in the South West than in any other region except the North. While these figures are for the South West as a whole, the major concentration of defence contractors is in the Bristol subregion itself, with the rest mainly in the broader 'defence manufacturing zone' extending from Yeovil in the south, through Bristol and Bath, to Cheltenham and Gloucester. Defence spending has, moreover, underpinned the continued development of, specifically, technological advanced manufacturing and R&D activity, with major export repercussions. Payments under defence contracts are made direct to firms, and frequently cover development costs and offer guaranteed markets. This contrasts with the much more broadly based activity funded by regional assistance. Support for civil projects has reinforced this. In 1972–3, for example, funding for Concorde alone, largely benefiting Bristol, was £245 million. Total spending on 'regional support and regeneration' for the country *as a whole* came to only £547 million. Other projects have also benefited under general support for R&D and the aerospace industry including, as we have seen, Hewlett Packard. While the locality has not therefore received regional assistance as such, it has been a major focus for central government expenditure, in particular through defence equipment procurement.

Local economic policies

Finally, local councils in the area, as elsewhere, have established explicit economic development measures. The major role has been played by Bristol District, with the other district councils playing an essentially supporting role and Avon County a late arrival on the scene. Through the long boom, and indeed up until the mid-1970s, the city council's main involvement in what would now be termed 'economic development activity' was through the development of serviced industrial estates—an activity which it pursued on a significant scale. Policy aims were to accommodate firms affected by war damage and later planned redevelopment, to maintain and diversify the manufacturing base, and to provide employment opportunities adjacent to large public housing developments in south and east Bristol. Shortage of industrial land in the later 1960s, and growing demand for warehousing and distribution developments, prompted the development of new estates at Avonmouth close to the port and motorway links. The local authority thus played a major role in the provision of sites for manufacturing industry building, comparable in scale with the part played by other authorities with a tradition of estate provision such as Sheffield and Derby (Boddy 1982).

The other significant development essentially rooted in the long boom and a vision of economic progress was a major investment in a new facilities south of the existing nineteenth-century facilities at Avonmouth (Bassett and Hoare 1984). This was first planned in the 1950s, the city-controlled Port Authority acquiring major land holdings for this purpose in 1959. Despite protracted debate through the 1960s, government financial assistance was refused on the grounds of anticipated traffic and commercial viability, and conflict with national policy to concentrate traffic on existing ports. The city nevertheless obtained parliamentary approval in 1971 for a Bill allowing development to proceed: this would be funded entirely by council borrowing, with the council committing itself to a massive public investment project. The plans received strong bipartisan support and backing from the unions and assorted business interests. This was apparently sustained by an ideology which could draw together a determination to carry forward the city's entrepreneurial mercantile heritage and a commitment to municipal investment on the grand scale, in the face of all opposition and, to some extent, reason. Costs had risen to £37 million by the time the dock opened in 1977, coinciding with deepening recession and a slump in trade which resulted in severe operating losses. By 1981–2 a deficit of £11.7 million had accumulated, imposing a massive burden on the city's finances. By 1984–5 the rates burden of the dock debt had risen towards £13 million. On the other hand, around 6 500 jobs were tied up in the dock and dock-related industry, and the total employment impact of the docks, together with wider multiplier effects, has been put as high as 15 000.[27] After a long series of debates and manœuvres, a programme of asset disposal, primarily of council-owned retail property in the city centre, was initiated to repay the outstanding debt of, by 1984, £55 million; however, the underlying economic problems of the docks remain. Furthermore, the opportunity costs of the financial resources committed to the project overall need to be taken into account, and were a backdrop to the development of more explicit economic development policies, narrowly defined, from the mid-1970s.

Manufacturing job loss accelerated in the early 1970s and unemployment started to increase, rising, symbolically, above the national rate. The council was also increasingly aware of possible competition from the expansion of economic initiatives by other local authorities; aerospace employment was starting to look unstable, particularly as Concorde ran down; and regional policy seemed to be a problem, with the loss of Toyota. Bristol District responded in 1976 by setting up a broad-based Economic Development Board for the whole Avon area with representatives from the constituent councils, unions, employers, and educational establishments, followed by an Economic Development Subcommittee of Bristol's Resources and Co-ordination Committee and the appointment of an Industrial Development Officer. Policy development was largely concensual despite a switch from Labour control to a period of no overall control in which (to simplify) first the Conservatives and the Labour established a majority with Liberal support. Changes internal to the Labour Party locally, however, saw the tentative emergence of more radical policy elements and some questioning of established directions. Moves by Avon County followed Labour's taking control in 1981; Labour retained control in 1985 with Liberal support. Policies were evolved in a number of directions. First, the preservation of existing industry in the face of manufacturing decline was a major objective. In practice, however, little positive could be done beyond attempting to avoid obvious planning obstacles and to assist with needs for land and premises. Financial assistance was provided in only one or two specific instances. Rate and harbour-due concesions to Rio Tinto Zinc, for example, may have had a marginal influence on its decision to continue smelting operations in the face of European-wide rationalization. Finally, the council joined in (unsuccessful) lobbying activity to pursuade British Leyland to retain a local plant employing 560 people.

Support for small firms was pursued, initially through action on sites and premises. Subsequently, two small-firms officers were jointly funded by the other districts for a two-year period from 1979 to offer aid and advice. By 1981, however, it was argued that private sector agencies could meet these needs and that resources could be more effectively used to fund a single more senior post with broader responsibilities together with an increased contribution to the promotional budget. The need was identified for small low-cost units and the provision of 'enterprise workshops' was initiated with a first development in St Pauls and a second in south Bristol. Four conventional developments of small units have also been completed. Most recently, reflecting shifts in Labour Party thinking, two Co-operative Development Officers were appointed to work with the Avon Co-operative Development Agency.

Traditional land assembly and estate development remained a priority. This included site availability for major incoming firms, particularly on the northern fringe where the city was able to provide sites for Hewlett Packard and Dupont, and to the west at Avonmouth where estates were expanded for warehousing and distribution. Concern over the mismatch between this pattern of land availability and job loss in east and south Bristol led to site development on the former board mill site and a partnership development with a private developer on a site adjacent to major council estates to the south. Site provision has been backed by the information service and property registers maintained by the Economic Development Office. Capital expenditure for the period 1977–8 to 1983–4 amounted to £7.2 million—about £2 milion for small industrial units, and the rest for site acquisition and infrastructure. Around £3 million of this capital expenditure went to the inner area, £2 million to south and east Bristol, £1 million to Avonmouth, and £1.5 million for the site purchased by Hewlett Packard. A further £3 million was budgeted for 1984–5, mainly for sites and infrastructure, including a further £1 million on the Hewlett Packard site.

Particularly prominent in policy terms, and accounting for much of the Economic Development Officer's efforts, however, has been promotional activity and action to attract in primarily overseas investment. The USA was seen as the major source with West Germany offering some potential, and the emphasis was placed on electronics, pharmaceuticals, cosmetics, and plastics (Bristol City Council 1978). Promotional and marketing activity has been backed up by regular visits and 'missions' to the USA, particularly to California and the Boston area, and agency services were established in 1982 through US consultants. These emphasized labour market characteristics, aerospace, communications and intrastructure, and quality-of-life factors together with the 'high-tech', M4 growth corridor image reinforced by publicity given, in particular, to the Hewlett Packard decision. More limited activity has been directed towards West Germany and Japan, but has included visits as well as promotional activity. Within the UK, a limited amount of promotional activity has been directed towards London and the South East. Thus, for 1984–5, the key targets were the USA and the Thames Valley (Bristol City Countil 1985). The particular emphasis on high technology has undoubtedly been fed by the inflated claims of job creation in electronics companies and a desire to 'talk up' the locational attractions of the area: in 1985, for example, the Economic Development Office was talking of 4000 new 'high-tech' jobs by the end of 1986, though this bore no relation to the expansion plans of existing companies or publicly announced plans for new establishments. No particular priority in promotional terms, on the other hand, has been given to attracting in financial services, despite the scale of spontaneous relocation from the early 1970s. The role of development companies and agents in advertising and promoting specific sites and premises, with total expenditures far exceeding those of the city council, coupled with the impact of more general reviews in the trade press, should be remembered, however.

Finally, increasing prominence has been given to the economic and employment potential of service-based commercial promotion and activities such as exhibitions, fairs, and conferences, tourism, and leisure. Much of this has been centred around the redevelopment of the old city docks area and historical and maritime themes, with support for a range of specific events including the World Wine Fair, Grand Prix power-boat races, and the construction of a Maritime History Museum alongside the historic SS *Great Britain*. Current proposals include promotion of additional conference, exhibition, and hotel facilities. This general policy area was consolidated in 1983 with the establishment of a Bristol Marketing Board, separate from the Economic Development Office and jointly funded by the city council and local firms. Further initiatives followed in 1984 together with the English Tourist Board, including the drawing up of a 'tourist development action plan', aiming in part to lever in further central government funding for historic conservation; such funding had amounted to nearly £2 million since 1977, and had encouraged private investment on top of this (Financial Times, 24 July 1984).

In 1982, Avon County itself set up an Economic Development Subcommittee of the council's Resources Committee with an initial budget of £1 million. An initial strategy was set out drawing on consultation with progressive Labour councils, including West Midlands County, and taking account of Bristol's existing activity. Reflecting specific political commitments within the ruling group, this emphasized support for small and medium-sized local firms, development of co-operatives, support for new technology and product development, financial provision linked possibly to planning agreements and an 'enterprise board', provision of land and premises including communal workshops, and a local purchasing policy. As elsewhere, faced with the problems and contradictions of implementing radical strategies in this area (Boddy 1984b), the reality was rather more traditional. By June 1984, direct expenditure totalled £274 000 and an essentially conventional programme of financial assistance had provided £100 000 in loans and just over £400 000 in guarantees. A development of small industrial units was under way, and Industrial Directory published, information again provided for small firms, and, in late 1983, a Co-operative Development Agency established. These developments were initially marked by, at times, open conflict and confrontation with Bristol City Council (also Labour controlled) rather than by co-operation. This reflected the spirit of antagonism rooted in Bristol's loss of status at reorganization when Avon was created, Bristol's established dominance and 'ownership' of economic development activity in the area, and the gap between traditional and more radical emergent elements within the Labour Party locally. Conversative controlled and essentially non-interventionist, the other districts in the area have had very limited involvement. Activity has extended to joint funding of the small-firms officers, financial support and occasional participation (in some cases) in Bristol's promotional activity, and a few small industrial developments including small units and workshops. Northavon District, covering the northern fringe where pressure for growth has been concentrated, has been rather more pro-development, actively promoting schemes primarily through the planning framework; it provided rate relief to Hewlett Packard, who located on land owned by Bristol but in Northavon District.

As already noted, the Bristol area had not until 1986 been granted Inner City 'Partnership' or 'Programme' status. It had, however, been one of the highest spenders under the 'traditional urban programme' through which other authorities bid for grant aid for individual projects. The Bristol and Avon area as a whole was allocated £460 000 for 1984–5, (75 per cent met by central government), covering social, environmental, and economic projects. Bristol Council had, however, argued repeatedly for full Programme status. It had sought to justify this on the basis indicated in the previous section, that high absolute levels of unemployment and related problems, and their localized intensity, particularly in St Pauls, are masked by aggregate statistics. Thus Sheffield and Coventry, as measured by the OPCS report (Redfern

1982) quoted earlier, had fewer people in problem wards than Bristol, and total unemployment in 1984 was roughly the same in all three areas. Sheffield and Coventry, however, both had Programme status, and both were allocated over £4 million in 1984–5, compared to the Bristol area's £460 thousand.

More speculatively, in 1984 Bristol argued the case for Assisted Area status to be granted to a contiguous area of twelve wards covering part of the inner area and the worst areas of south Bristol, with a population of around 200 000 and an overall unemployment rate of 20 per cent. The report noted that this area, with a population bigger than many towns in the UK, would, if it stood alone, 'without doubt, be one of the most depressed in the country' (Bristol City Council, 1984b). Further manufacturing losses were anticipated and the area was isolated from the northern fringe growth area. This bid, together with a further request for Inner City Programme status, was turned down. The evidence did, however, support a successful bid to the European Social Fund, and £196 000 was allocated in August 1984 to finance training projects for women, disabled people, and minority ethnic groups. In practice, little of the money—which represented something of a windfall—could be spent. This reflected the fact that the authority lacked the capacity for effective implementation within the time-scale which was imposed.

The scale and direction of economic development activity at the local authority level has not been exceptional in Bristol compared with many areas. The scale of expenditure is indeed relatively modest compared with that of authorities in inner London, the Midlands, and the regions, where, in many cases, both county and district councils have been heavily involved. Avon's low profile contrasts with county-level activity elsewhere, particularly in the metropolitan areas where capital expenditure and revenue spending, even under general 'Section 137' powers let alone the Inner Urban Areas Act, have been a major boost to overall economic development expenditure. Capital spending budgeted under the 'economic development' heading for 1984–5 was £3 million, while revenue spending was around £1.5 million, including £250 000 in Urban Programme grants from central government and £570 000 related to estate development. The budget for promotion and tourist development as such amounted to £180 000. The make-up of activity remains, also, essentially traditional, emphasizing provision of serviced sites and small units, promotional activity, and some support for small firms. Financial assistance—significant in many authorities—has, until Avon's more recent initiatives, been virtually non-existent and remains low key. Site provision, moreover, represents essentially the continuation and evolution of established 'estates' activity on a self-financing basis.

Local authority estates have represented a major part of the supply of serviced sites in the locality. Provision has obviously facilitated economic growth and efficiency, although its precise impact cannot readily be assessed. This turns on the perennial question of how far local authority activity was displacing or complementing private sector development activity. Local authority reports, at least, suggest that the council was responding to perceived needs and, at times, recognizable shortages rather than undertaking development for its own sake. Site provision has complemented the private sector's greater willingness to become involved in provision of premises. In specific cases the local authorities have been prepared to become heavily involved with a package of measures, Hewlett Packard again being the main example. The city council provided a serviced site on advantageous terms including the option of phased purchase by the company of up to 165 acres at a fixed price, providing it with both flexibility and a hedge against inflation. The county provided rates relief worth £180 000 and, with Bristol, contributed £300 000 to the cost of infrastructure. This was supplemented by central government's £5 million R&D grant. There *are* other specific examples of innovative and effective initiatives, initiated or supported by the local authorities, with localized impacts, but they remain essentially insignificant in wider economic and employment terms.

The specific impact of promotional activity in terms of relocation, employment growth, and effects on different labour market groups and areas is particularly difficult to disentangle from the overall process of locational and investment decisions. Effects are likely in many cases to be indirect and essentially unverifiable. For this reason, evaluation, in Bristol as elsewhere, is commonly based on numbers of firms contacted, enquiries received, and firms locating in the area, rather than causal connections. Specific claims have drawn largely on the Hewlett Packard experience, in which the Economic Development Office does seem to be heavily implicated. In other cases, including the relocation of insurance companies such as London Life, that are linked with Hewlett Packard in publicity material, the connection has been less direct. The office has not so much influenced the choice of location, but facilitated the move and helped to sell the locality to key staff after management had identified Bristol as its preferred option. No significant inward investment has been generated by promotion targeted on Japan or West Germany. Apart from occasional instances where specific effects can be identified, promotional activity probably has to be based largely on faith, budgetary restraint, and attempts to maximize effectiveness. Nevertheless, the subcommittee concluded in 1985 that 'Inward investment from North America still remains the best area for creating new employment opportunities for the Bristol area' (Bristol City Council 1985). An exercise conducted for the city council in 1985 with site selectors in the USA, for example, suggested the need to focus on firms known to be considering locations within the UK and to promote the overall M4 Corridor rather than pursuing general promotional activity. More generally, the overall emphasis in economic development policies has been essentially on aggregate economic growth. This has been reflected in the priority given to provision of sites and premises, infrastructure, support for small firms, and stimulation of commerce and tourism. Distributional effects have, implicitly, been seen in terms of 'trickle-down' and multiplier effects through which the benefits of growth and prosperity are assumed to reverberate through the wider economic structure and labour market. Targeting of activity in terms of sectors or labour market groups has been relatively unsophisticated, reflecting the growth priority and a bias towards 'high-technology' and manufacturing industry. Attention to commerce and tourism seems to reflect its perceived economic impacts rather than any more general recognition of the importance of employment growth in the service sector. The employment structure of different sectors in terms of, for example, gender of skill has not been an issue, despite major differences between, say, electronics and insurance. Efforts have been made to ensure availability of sites and premises in areas of higher unemployment in east and south Bristol. Again, however, this in effect provides the preconditions for private investment decisions, coupled with a belief that locating industry in a particular neighbourhood will impact on unemployment levels within it. The location of an electronics company in south Bristol would doubtless be claimed as a major achievement. It would, however, be unlikely to have any greater effect on local unemployment than the Aztec West and Hewlett Packard developments have had on unemployment blackspots in the north. There has been much less concern with explicit distributive issues, directly linking job retention or creation to particular groups in the labour market. This had been limited to enterprise workshops and involvement in a number of small-scale training initiatives including an Information Technology Centre established with the Manpower Services Commission (MSC) and the Department of Industry. Nor has there been any real challenge to essentially private enterprise and conventional forms of production, as attempted in more progressive Labour authorities including the West Midlands, other than embryonic debates in some sections of the Labour Party and limited initiatives to support co-operative development.

The wider local authority role

Overshadowing economic development spending as such, and indeed much of the city council's activities, has been the docks debt. The original investment represented municipal investment on the grand scale. It is easier of course with hindsight to see the risks of such a strategy, and the background has been reviewed in detail elsewhere (Bassett and Hoare 1984). Dock development was not in any sense selected as one of a range of possible economic development or employment strategies, although the general economic impact on the subregion obviously to some extent lay behind the support it generated. The Port Authority at one point argued that 'Bristol's only hope for a healthy economy in the future lies in a major modernisation of its port' (Bassett and Hoare 1984, 226). This argument was not formally developed to any great extent, however, and indeed there was evidence even in the 1960s that subregional growth had little to do with the volume of port traffic. Although in many senses unique, Bristol's experience indicates the dangers of high-risk narrowly focused strategies that are particularly vulnerable to external forces. It emphasizes the need to identify specific objectives as a basis for developing cost-effective solutions—one can only speculate on the opportunity costs of the project, and what might have been achieved in terms of economic development and employment objectives by alternative uses of this volume of capital funds and rates revenue, exceeding the most ambitious plans for local enterprise boards.

Apart from specific economic development strategies, local authority policies and expenditure are implicated at a more general level. A generally permissive and development-oriented planning environment has been maintained, and this is reflected in survey evidence. A number of firms, including Hewlett Packard and London Life, remarked on the amenability and co-operation of the relevant authorities. This has amounted, not to a willingness to overturn existing policies, but rather to a willingness to respond to specific development proposals. Development on the northern fringe has been guided by specific development briefs, while the fact that the County Structure Plan was submitted much later than most may have given some element of flexibility. Office development has similarly been unaffected by constraint: local pressure led to a temporary freeze on new permissions for development in 1973 but the ban was meaningless in the context of major over-provision and given the scale of development under construction or already having permission. More generally, local authority planning and other policies have set the context and, in some cases, played a more specific role in terms of conservation, redevelopment, and environmental improvement. The abandonment of ambitious road schemes in the early 1970s, in the face of growing opposition from the mainly middle-class conservation lobby, preserved the urban fabric in a way which change in professional and public attitudes and financial constraints came too late to do in many cities. Subsequent local plans, the detailed implementation of planning policy, and varying involvement in a wide range of conservation and development schemes have contributed to the creation of an attractive 'urban milieu' and positive image. This, as we have seen, has played a specific part in locational terms.

As elsewhere, the major economic and employment impacts of the local authorities have of course been through their own mainstream expenditure and as employers in their own right. Local authority employment in the area as a whole fell by 12 per cent over the decade to 1981. There has also been a major drop in capital expenditure on housing and infrastructure in particular, reflecting central government's squeeze on local authority spending since the mid-1970s. Total current spending by the county and district councils together was in fact the same in real terms in 1984–5 as in 1978–9. Here the effects of central government's financial squeeze (Boddy 1984a) have been more subtle, but none the less damaging. For, though spending as a whole has not fallen, cuts in government grant have shifted the burden of expenditure much more heavily onto local ratepayers. The proportion of total expenditure met by government grant fell from 60 per cent in 1978–9 to only 44 per cent by 1984–5. This in effect represents a direct loss to the locality of around £160 million over the six years.[28]

The policy context: initial conclusions

The role of public policy is taken up more generally in the following section, but a number of initial conclusions can be drawn. First, in terms of explicit urban and regional policy, the locality has, on the face of it, clearly been disadvantaged in terms of financial resources compared with the regions and inner cities. The negative impacts in broader terms of economic performance and employment are harder to pin down. There are specific instances of IDC refusal and the Inmos example. More general studies of the effectiveness of regional policy, however, do not suggest that it has specifically diverted a significant volume of manufacturing industry to the regions that would otherwise have located specifically in the Bristol area. Regional policy has in any case grown weaker since the late 1970s. Office dispersal policies, as opposed to manufacturing location policy, appear in fact to have worked to the area's benefit, the scale of office development representing a significant factor in service sector growth. Inner City funds have been positively linked to economic and employment gains, but on a very limited scale in absolute terms.[29] Even this has been highly dependent on how funds were deployed and whether they replaced main programme spending. Negative impacts in terms of explicit urban and regional policy have been more than outweighed in economic and employment terms by the other government programmes and expenditure—particularly defence spending and support for aerospace and, to some extent, nationally planned transport infrastructure. Locational attributes specific to the area were identified earlier, including access and communications, labour market structure, the urban environment and quality-of-life factors, and, finally, availability of land and premises. This implies, in particular, the importance of infrastructure provision, support for aerospace and its influence on labour supply and the area's skill reputation, and the lack of restriction on office development.

Local authority initiatives, whatever their individual successes, have had little overall impact relative to the scale of economic and employment change. The facilitative role played by site provision may well have been more important than specific promotional activity. Specific to the Bristol locality, major investment in dock facilities resulted in major financial problems for the council and has been significant mainly in terms of lost potential for economic and employment initiatives, had equivalent resources been made available to this end. Finally, cuts in local authority employment and the financial burden imposed by withdrawal of central government grant have been of much greater relevance than specific economic or employment initiatives at the local level. The specific locational attributes of the area suggest the importance, in particular, of the local authorities' role in maintaining and enhancing the quality of the urban environment and amenities, the provision of industrial sites, and the operation of an amenable planning regime.

Discussion

Recession and rising unemployment by no means bypassed the Bristol subregion. Compared with urban areas beyond the prosperous South, however, it has survived relatively well. As we have seen, manufacturing job loss hit some sectors hard. This, however, left a core of more stable employment in traditionally dominant sectors. Aerospace, moreover, shored up manufacturing employment, maintaining a dynamic core of technologically advanced activity with wider economic and employment repercussions. Particularly important to aerospace has been government defence spending and, to a lesser extent, support for civil aerospace production. Lastly, despite manufacturing job loss, Bristol has consolidated its role in terms of corporate hierarchy, retaining and extending R&D and higher-order management functions. This final section draws out some of the main conclusions and implications of the analysis in relation to economic change, the impacts of change, and policy implications.

Economic change

Development of technologically advanced activity, including electronics-based R&D and production, is heavily concentrated in the long-established aerospace sector, much of it specifically defence related. High-technology growth, in the sense of an influx of rapidly expanding electronics and computer companies or the mushrooming of new enterprises, has been relatively unimportant in economic or employment terms. One can find examples of such companies. Bristol's 'high-tech' growth image and Silicon Valley comparisons are, however, essentially wishful thinking. Nor do employment growth prospects in the electronics industry and Bristol's likely share of national expansion suggest that this is likely to change dramatically in the future. In fact, analyses of high-technology activity in the USA in particular[30] have emphasized this centrality of defence and defence-related R&D and production in high-technology employment (Saxenian 1980; Glasmeier *et al.* 1983; Malecki 1984; Markusen 1984a). This in turn, it has been suggested, has formed the basis for a specific spatial pattern of high-technology-based development and urban growth (Castells 1983; Markusen 1984b). The Bristol case emphasizes the extent to which such defence-related high technology is implicated in regional economic performance and the development of specific urban areas.

As elsewhere, service sector growth has offset manufacturing job loss. Manufacturing will soon account for little more than a quarter of all employment. The extent to which urban areas attract or generate service growth, and the structure of that activity, will be a crucial determinant of economic prosperity and the scale and pattern of employment opportunities. For the USA, for example, Stanback (1979) has shown how metropolitan areas in regions favoured by growth and already specialized in service provision have experienced the strongest growth in business services, suggesting significant parallels with the Bristol case. As the Bristol example demonstrates, however, it is particularly important to disaggregate overall 'service employment' in terms of processes of change, labour market implications, and the way in which different elements of service activity articulate with economic activity in general. Locally, while the overall shift from manufacturing to service employment reflected national trends, growth was concentrated in two areas in particular. The first was financial services. This reflected the 'relocation appeal' of the locality for 'basic' activities exporting services beyond the area, but also its increasing nodality as a regional financial and business centre.[31] Scale effects and the clustering of additional financial and business services in the urban area seem to have followed from this overlaying of major regional functions with basic activity. Growth of office-based services both stimulated and was facilitated by the boom in commercial property development, reflected in construction and related employment. Again, Bristol's role in terms of corporate hierarchy and function has been consolidated with the growth of headquarters functions and the centralization of regional-level activity.

The second growth focus has been 'Miscellaneous services', including the 'eating, drinking, and entertainment' services but also the mixed bag of other services, not broken down by official statistics, all of which expanded much faster than nationally. Commonly seen as 'dependent' on demand generated elsewhere in the local economy, the strong growth of these activities reflects the strength of other sectors in the locality. It seems to reflect more specifically the relative importance of higher-income professional, technical, and white-collar staff, concentrated for example in aerospace, electronics, and financial services. Overall, the local situation supports the sense, if not the detail, of Hall's suggestion that:

> the growth of a relatively small industrial base especially in innovative, high-technology industry and associated producer services—can create a very large income and employment multiplier effect in the form of construction, real estate, recreation and personal service industries. (Hall 1985, 10.)

The importance of 'basic' services and the role of the local 'business services infrastructure' has increasingly been seen as a positive factor in stimulating economic activity at the urban or regional level (Marquand 1979). A range of personal and miscellaneous services would also, however, appear to be, in the Bristol case, an important component of the image and attractiveness of the locality, both to corporate decision-makers and to particular types of labour. Beyond these sectors, a large part of past service employment growth and, more recently in some sectors, decline has been dependent on public expenditure. Future employment levels are particularly dependent on public policy and expenditure decisions—potentially much more so in the current political and economic climate than on the impacts of new technology. Shrinkage of public service employment could thus offset much of any employment gain in other sectors. Overall, however, the case of Bristol supports Stanback's conclusions, mentioned earlier, as to the importance of service growth in the better-than-average adaptation of particular urban areas.

The picture of economic change, although generally favourable compared with places like Glasgow, Newcastle, and the West Midlands, has by no means been one of successful restructuring generating significant job opportunities across the board. Growth in some sectors has been juxtaposed with decline in others, and the broad pattern of change has been determined as much by how well existing employment has stood up in the face of decline as by which sectors have been expanding. Bristol's 'success', particularly in manufacturing, partly reflects the fact that employment locally has simply declined less than elsewhere. Growth as such has, as we have seen, largely been confined to particular components of the service sector. This reflects the fact that the geography of job change, more generally, has since the mid-1970s been largely a question of how well existing employment has stood up in the face of decline—how far closure and cut-backs in surviving enterprises, coupled in some cases with reinvestment, has hit different sectors. Overlaid on this has been a limited amount of relocation and new investment directed to particular localities.

The performance of the Bristol area itself reflects both a number of specific factors and the particular legacy bequeathed by previous rounds of investment. It is a major urban area with an urban and subregional environment, 'quality of life', and social image attractive to corporate decision-makers and to elite groups in the labour market. The labour market itself is diverse, including, for example, high-level R&D, technical, and professional staff, skilled manual workers with experience of advanced manufacturing techniques, good-quality school-leavers, and basic assembly or routine clerical workers. It has built up major regional service and administrative functions within a generally prosperous wider region. Finally, it is particularly accessible to London and more generally, yet divorced from the metropolitan labour and property markets. Added to these specific factors, the legacy of previous rounds of investment was in many ways favourable. The area's traditional manufacturing sectors declined later and less severely than the manufacturing core in many localities and have been consolidated, albeit on a reduced scale. Aerospace, moreover, gave the area a major stake in one of the few manufacturing sectors to survive relatively well in employment terms since the long boom of the 1950s and 1960s. This legacy was reflected in the scale of job loss and the area's favourable performance (particularly since the late 1970s), in the favourable structure of the labour market already noted, and in the lack of large-scale physical dereliction associated with traditional heavy manufacturing. Looking to the future, while the Bristol area is well placed to capture a significant share of any future growth, the scale of unemployment and the likely balance between net new entrants to the labour market is such that job growth in the short or medium term is unlikely to take up much of the slack in the labour market overall. Certain segments of the labour market are likely to overheat fairly rapidly. Reduction in unemployment as a whole, even under optimistic forecasts, is unlikely to be rapid or far reaching. In a relatively prosperous locality, as in the country as a whole, then, return to 'full employment' based simply on economic recovery is not a realistic possibility.

The 'good city' for everyone[32]

The overall picture in terms of unemployment and job loss compares favourably with that of cities in the Midlands and North. It is obvious, however, that the benefits of this are not evenly distributed through different groups in the labour market. This is emphasized by the scale and increase in youth and long-term unemployment. In part, this reflects the pattern of sectoral change. Male manufacturing job loss has been juxtaposed with increasing female service sector employment, much of it part time. Job opportunities, beyond those created by growth primarily in parts of the service sector, have been dependent on replacement for turnover of existing staff. Growth in particular has created few job opportunities for the majority of less qualified school and YTS leavers and for many of the semi-skilled and unskilled 'shaken out' of employment. Recruitment, generally, has emphasized skills, experience, and qualifications which exclude the majority of these groups. In particular, in many major employers there has been little recruitment from the unemployed, the majority of vacancies being filled by those in employment, with skills and experience. In the context of employment shrinkage, vacancies created by such moves will often be left unfilled. The main exceptions have included, for example, miscellaneous services, retailing, typing, and junior clerical occupations.

These processes have been reflected in increasing polarization between core and peripheral groups in the labour market. Core groups include the relatively well paid and securely employed with relevant skills, technical knowledge, or professional expertise, together with those possessing the qualifications who can successfully compete to join them. They are likely to enjoy the prospect of career or salary progression and the possibility of job mobility with little risk of unemployment, certainly for any length of time. Union structures remain important in much of manufacturing and public sector employment while other groups benefit from relatively benign management–labour relations based on the enlistment of co-operation and consent, and a non-conflictual technical and professional ethos. Employment is disproportionately white and male. Peripheral groups include the semi-skilled and unskilled, the less well-qualified school-leavers, and those whose employment or employment prospects are tied to routinized, de-skilled work in manufacturing. Also included are parts of the expanding service sector offering relatively low-paid and insecure employment, possibly part time and commonly non-unionized office and clerical work with little prospect of career advancement—groups that have been termed 'service workers' as distinct from white-collar career grades and professionals. Peripheral groups are characterized by a much greater concentration of black workers and, particularly in services, women.

Polarization in part reflects simply the pattern of job gain and loss and its impact on different groups in the labour market. Core employment has been maintained by the resilience of certain sectors, while peripheral employment has grown in particular with service sector expansion. Stability and growth of core labour in parts of manufacturing and services are, however, structurally linked to a dependent growth of peripheral employment, much of it female and part time, in the service sector, particularly miscellaneous services. The relationship between basic and dependent activity and the operation of what have been called 'trickle-down' effects are thus reflected in structured inequalities in the labour market. There is also evidence both locally and nationally of increasing differentiation within particular firms and sectors, between those in occupations where skill, specialization, and technical content are increasing, linked in some cases to new technology and with career prospects, and those performing tasks such as routine assembly, data inputting, and basic clerical work. This overall process of polarization is likely to continue. In part this reflects the continuing shift from manufacturing to service employment. The latter tends overall to have a higher proportion of lower-paid workers, women and black people, part-time work, and less secure employment. Technical change is, moreover, likely to increase polarization within specific firms and sectors in both manufacturing and services. Particularly relevant to Bristol, Rajan (1984) predicts increasing differentiation within financial services between 'career' and 'non-career' clerical staff. In the UK electronics industry, Soete and Dosi (1983) see job gain in high-skill categories but significant loss of jobs affecting operators and, to some extent, clerical and craft workers, with female employment most affected by decline. Within more dynamic sectors such as electronics, aerospace, and insurance, economic recovery will, moreover, be characterized by 'jobless growth' in output with less than proportional employment increase. In terms of services as a whole, the Bristol case parallels Stanback's suggestion from the US experience that

> we are moving towards a sharply dichotomized service work force offering, on the one hand the skilled, responsible, and relatively well-paying jobs of certain professionals, trained technicians, or artisans, but on the other, the unskilled, undemanding, and poorly paid jobs of salespersons, service workers, or laborers. (Stanback 1979,106.)

If, as suggested earlier, Bristol is indeed at the forefront of more generally emerging processes of change, then these observations are obviously of wider relevance.

Policy implications?

The wider policy implications of the Bristol case turn both on the extent to which specific policy measures are implicated in the observed patterns of change and on the extent to which these are relevant or reproducible elsewhere. In naïve terms of direct and practicable policy lessons applicable more widely, the conclusions are limited. In part this is simply because Bristol is no more divorced from the working out of national- and international-scale processes in a particular unique locality. The complex legacy of historical processes which constitute this particular locality cannot in any simple sense be reproduced. As we have seen, no specific urban or regional policy measures, central or local, are heavily implicated in the particular processes of economic and employment change which have been set out. There are examples of innovative and, at the local scale, effective initiatives from which lessons can be drawn. Their impact in overall economic and employment terms has, however, been essentially irrelevant. Location, environment, quality of life, regional centrality, labour market structure, and, in particular, the legacy of past rounds of economic activity are among the key explanatory factors—and among the least tractable in policy terms. The importance of image, urban environment and 'milieu', and quality-of-life factors emphasizes the importance of policies and expenditure which bear on these.

The strength of miscellaneous services and the possible importance of the personal service sector as an attractive factor suggest, the role of measures to stimulate personal and business tourism, leisure, and recreational activities. This has, however, particular employment implications, including the growth of relatively low-paid part-time female employment. Together with the importance of financial services, Bristol's experience, nevertheless, emphasizes the importance of developing explicit policies relating to the service sector and breaking spurious and outdated obsessions with manufacturing as the only source of 'real' jobs. Service growth locally cannot, however, be divorced from the specific character of the Bristol area and the 'raw material' this afforded. Finally, the importance of site and premises availability to suit specific users and the ability of the local authorities and other agencies to respond positively have been emphasized in specific instances. Again, however, these are merely necessary factors, reproducible and indeed duplicated in many localities, and far from sufficient to attract economic growth. It is in fact precisely the absence of explicit central government urban and regional policy measures which characterizes the area. As suggested earlier, however, while the area has clearly lost out in straight financial terms from lack of Urban Programme funds and regional assistance, this does not appear to have been a major disadvantage in overall economic and employment terms. Office dispersal policies in fact worked to Bristol's advantage and, with hindsight, suggest misconception in the way these policies were applied.

In terms of specific Inner City policy, Bristol's problems are clearly less severe than those in many localities. The absolute scale of unemployment and related adverse indicators are, however, masked by authority-wide averages and their multi-locational character. They display the increasing double-sided nature of 'inner city' problems including both older inner areas of mixed housing tenure and peripheral council estates. On the latter, escalating housing and environmental problems, coupled with the continuing link between the poorest housing and the worst-off households, relate in particular to local authorities' financial capacity. This has been particularly limited in the present political and economic climate. Urban Programme spending is minimal in the face of the scale of the economic, social, and environmental problems it confronts. The Bristol case supports, nevertheless, the need for more finely focused targeting of what funds are made available. In this situation, parts of Bristol would clearly qualify for additional spending.

Much more heavily implicated in the locality's relative prosperity is, as we have seen, the spatial impact of government defence procurement and support for aerospace. The relationship between the defence sector and the national economy and the character of defence product markets has received considerable attention. In this country at least, however, there has been little explicit attention paid to the spatial economic or employment dimensions of defence procurement or sectoral industrial support beyond explicit regional assistance. There has been some concern over the employment effects of naval dockyards and shipbuilding in the Assisted Regions and localized lobbying activity. Policy concern with the spatial impact of procurement as such has been insignificant.[33] Awareness, more generally, of the spatial distribution and effects of defence spending has increased both in the USA, in debates around high technology (noted above), and in relation to the UK (Short 1981; Law 1983). Some of the issues are discussed in more detail elsewhere (Lovering 1985; Boddy and Lovering 1986). In the USA, it has been argued (Markusen 1984a), military spending in the post-war period, has increasingly served as an implicit industrial policy, spatially oriented to the least depressed, 'sunbelt' areas outside of the old manufacturing core. Similarly, in Canada, the federal government has played a key role in the uneven spatial development of aerospace which has been the subject of explicit political conflict (Todd and Simpson 1985). The Bristol case emphasizes for the UK the impacts of such 'implicit spatial economic policies' which have operated, particularly through defence procurement, but also through other forms of sectoral assistance. The pattern of such expenditure has, as we have seen, largely run counter to explicit regional and urban policy, and explicitly underpins technologically advanced R&D and production. These spatial economic and employment impacts, and the scope for explicit spatial policy inputs, are undoubtedly complex and raise a range of issues. Spatial policy concerns, for example, are obviously traded off against cost and efficiency and, possibly, strategic factors. Clearly, however, it is important that these issues are explored. Such expenditures are obviously heavily implicated in the pattern of urban and regional economic performance, and their impacts must be explicitly considered in policy terms. More generally, this issue indicates the need to look closely, not just at explicit urban and regional policy measures, but at the spatial impacts of the full range of state policy and expenditures if we are to understand spatially differentiated processes of economic and employment change, and the specific role of the state in relation to different localities.

To the extent that Bristol has 'succeeded', it has done so largely, in a sense, at the expense of other localities. This emphasizes the competitive nature of the policy issues raised. In a crude sense, insurance growth, for example, was achieved at the expense of London. Had it been possible to attract the companies involved to Wales or Glasgow they would obviously then have been lost to Bristol. The result would have been similar, had aerospace projects been allocated to the companies' other sites in Derby or Lancashire, or if Nottingham rather than Bristol had been selected as the headquarters of Imperial Tobacco Limited. This is essentially a truism, of course, but it emphasizes the fact that, in terms of the conscious spatial allocation of investment and employment by enterprises or by the effects of spatially directed policies, Bristol's relative success does not reflect some intrinsic growth capability. It is the other side of the coin to relative failure elsewhere. There are exceptions: Hewlett Packard, for example, was choosing between Bristol and Lyons in France so that Bristol was not in competiton for jobs, at least with other UK localities. Other companies have expanded from local roots, but in limited numbers.

The uneven distribution of the benefits of Bristol's economic performance and the tendency towards increasing polarization within the locality emphasize that economic growth and policies directed to this end have specific and unequal distributional implications. Regional and urban policy and, to a large extent, local authority policies have been primarily oriented to growth rather than redistribution. The goals of such policies as far as many localities are concerned are in effect to emulate Bristol's 'success'. Without explicit concern with the distribution of 'success', such policies, to the extent that they succeed, are destined to reproduce the sort of uneven and polarized development evident in the Bristol case.

Acknowledgements

The study reported here was funded by the ESRC, grant numbers DO 320005 and DO 3250012. The project relied heavily on intensive interview-based information gathering. We would like, therefore, to thank the many individuals in the companies, local authorities, employment and training agencies, and other organizations in the locality who were involved in this process.

Notes

1. *The Economist*, 30 Jan. 1982.
2. *Sunday Times*, 26 July 1981.
3. Interviews were conducted with thirty employers in five key sectors. The sectors were selected on the basis of their importance to the local economy and their contribution to economic and employment change. They included Tobacco, Paper and packaging, Aerospace, Electronics, and Insurance, defined in terms of Annual Census of Employment Minimum List Headings (MLHs)(1968). The primary study area, illustrated in Figure 2.1, is the Bristol 'travel-to-work area' as defined by the Department of Employment. Unless otherwise noted, this is the area or locality referred to as 'Bristol'. It roughly coincides with Avon County, less Bath District to the east and Weston-super-Mare to the west.
4. A much fuller account and analysis is presented in Boddy *et al.* (1986). Detailed studies focusing on the five individual sectors have been published by the School for Advanced Urban Studies, University of Bristol. Other, more specific analyses include: Bassett (1984); Bassett and Hoare (1984); Lovering (1985, 1986); Boddy and Lovering (1986).
5. Following Department of Employment convention, 'manufacturing' is defined as 1968 Standard Industrial Classifications (SICs) III–XIX and 'services' as SICs XXII–XXVII. Services thus excludes the public utilities, Gas, water, and electricity, included in some definitions.
6. This breakdown is analysed in more detail in Boddy *et al.* (1986), ch. 2. Predominantly public services include MLHs 901, 906, 701, 702, 706–9, 872, 874. Predominantly private services include MLHs 703–5, 709, 810–12, 820–1, 831–2, 860–6, 871, 873, 875–6, 879, 881–9, 892–5, 899.
7. Figures are from public expenditure White Papers, various years, converted to constant '1980 Survey' prices.
8. Includes MLHs 272, 354, 364–7, 383. For more detailed discussion of definitions of high technology, see Langridge (1983) and Boddy and Lovering (1986). The latter includes an extended analysis of the structure and role of high technology and aerospace in the Bristol locality.

9. Information from *Annual Reports to Council*, 1983–4 and 1982–3, University of Bristol, Appendix C.
10. Prime office rents in Bristol in 1973–4 were £2.00 per square foot, compared with £16.00 in the City of London and £12.00 in Westminster. By 1983–4, taking office rents and rates together, office costs in Bristol were around £8.10 per square foot compared with £47.70 in the City of London, £29.60 in Westminster, £14.90 in Reading, and £13.70 in Croydon (Debenham, Tewson, and Chinnocks 1983). They were, however, no higher than in other more distant provincial centres such as Manchester, Sheffield, or Leeds, so that there was nothing to be gained in this respect by moving further from London.
11. Figure for Avon County as a whole (source: Avon County Council 1985).
12. Information from Department of Trade and Industry, Establishment File, special tabulation.
13. Meaning outside of the Bristol travel-to-work area.
14. Figure for Avon County (source: Avon County Council 1985).
15. Figure for an area approximating the travel-to-work area comprising Avon County minus Bath District and the Weston-super-Mare sub-area of Woodspring (source: Avon County Council 1985).
16. Calculated by Begg *et al.* (1985). Bristol was one of only eight cities out of the six conurbations and seventeen largest urban areas not to have lost employment over this period. The others were Coventry, Cardiff, Plymouth, Derby, Southampton, Aberdeen, and Portsmouth.
17. Figures for employment (from Census of Population, Work-place Tables), as opposed to resident work-force, indicate the same, with employment falling in Bristol District by 3 per cent and rising in the rest of the county by 8 per cent.
18. Avon County Council (1985).
19. Female unemployment is not discussed here, given the inadequacy of published information based on numbers registering as unemployed or claimants as a true measure of numbers of women seeking, but unable to find, work.
20. Department of Employment, special tabulation for travel-to-work area.
21. Avon County Council (1984).
22. Female minority ethnic group unemployment was significantly lower than female unemployment as a whole—4.9 per cent compared with 8.1 per cent. This, however, reflects in particular the inadequacies of conventional official measures of female unemployment (see note 19).
23. Avon County Council (1985).
24. Bristol City Council (1984a).
25. Comparing the highest-ranked zones in terms of unemployment in 1971 and 1981. There is some shift of the worst areas into east and south Bristol, but only marginally.
26. Quoted in Ludlow (1983, 56).
27. MSC, Employment Division, Area Office, January 1981.
28. Analysis based on 'Financial, General and Rating Statistics', Chartered Institute of Public Finance and Accountancy. Converted to real terms using the gross domestic product deflator, *Economic Trends*, Central Statistical Office, HMSO. Loss indicates the sum which Avon County and Districts would have received in rate-support and block grant had annual expenditure been as it was and if the proportion of expenditure met by grant had remained constant from 1978–9.

29. A recent survey of 504 firms which had received Urban Programme funding, and which employed 9000 full-time equivalents, suggested that Urban Programme funds were specifically responsible for the creation of just over 400 jobs and the retention of 136. The lack of such funding to a specific area such as Bristol is obviously irrelevant based on this (JURUE 1985).
30. As opposed to south-east Asia, predominantly specialized in the manufacture of bulk components, consumer, and some business electronics.
31. The Bristol and West Building Society, as if to reinforce this, explicity advertized itself in 1984 as 'Exporting to the rest of Britain' (*Investors Chronicle*, 2–8 Mar. 1984).
32. The term 'The Good City' is from the book of the same title in which the authors claim that 'All social groups and classes tend to benefit from the prosperity of the growing towns which depend more heavily on service industries, public and private, and the newer forms of manufacturing . . . The growing, prosperous city, the city which is kindest to its more vulnerable citizens and the city which distributes its opportunities more equally than most can all be the same place.' (Donnison and Soto 1980, 175–6).
33. The government's Special Preference Scheme was designed to allow producers in Development Areas to re-tender for contracts. Few contracts were, however, placed in the Development Areas under the scheme (Short 1981, 104).

References

Avon County Council (1984), 'The "Ethnic Minority" Population And Workforce of Avon and the Bristol Area', *Report to the Employment Working Group*, Planning Department.

—— (1985), *Draft Economic Review*, Planning Department.

Bassett, K. (1984), 'Coroporate Structure and Corporate Change in a Local Economy: The Case of Bristol', *Environment and Planning, A*, 16, 879–900.

—— and Hoare, A. (1984), 'Bristol and the Saga of Royal Portbury: A Case Study in Local Politics and Municipal Enterprise', *Political Geography Quarterly*, 3, 223–50.

Begg, I., Moore, B., and Rhodes, J. (1985), 'Economic and Social Change in Urban Britain and the Inner Cities', unpublished report, Economic and Social Research Council, Inner City in Context Initiative.

Boddy, M. (1982) 'Local Government and Industrial Development', *Occasional Paper* 7, School for Advanced Urban Studies, University of Bristol.

—— (1984a), 'Local Councils and the Financial Squeeze', in M. Boddy and C. Fudge, (eds.), *Local Socialism?*, London: Macmillan, 215–41.

—— (1984b), 'Local Economic and Employment Strategies, in M. Boddy and C. Fudge (eds.), *Local Socialism?*, London: Macmillan, 160–91.

—— and Lovering, J. (1986), 'High Technology Industry in the Bristol Sub-Region: The Aerospace/Defence Nexus', *Regional Studies*, 20, 217–31.

—— —— and Bassett, K. (1986), *Sunbelt City? A Study of Economic Change in Britain's M4 Growth Corridor*, Oxford: Clarendon Press.

Breheny, M. (1986), 'Contacts and Contracts: Defence Procurement and Local Economic Development in Britain', paper delivered to Anglo-American Workshop on High Technology Industry, Churchill College, Cambridge, mimeo.

Bristol City Council (1967), 'Manufacturing Industry and Warehousing in Bristol since 1945', Research Section, City Engineer and Planning Officer's Department.
—— (1978), *Annual Report*, Economic Development Joint Sub-Committee.
—— (1984a), 'Indicators of Deprivation in Bristol', *Research Report*, Planning Department.
—— (1984b), 'Statement of Case for Development Area Status'.
—— (1985), 'Annual Review of Economic and Employment Initiatives, 1984/85', Economic Development Joint Sub-Committee.
Castells, M. (1983), 'Towards the Informational City? High Technology, Economic Change and Spatial Structure: Some Exploratory Hypotheses', *Working Paper* 430, University of California, Institute of Urban and Regional Development, Berkeley.
Debenham, Tewson, and Chinnocks (Chartered Surveyors) (1983), 'Office Rents and Rates, 1973–1983', London.
Dodd, J., Bowden, R., and Whiting, M. (1983), *Defence Spending and the Electronics Industry*, London: Fielding, Newson-Smith and Co. (Stockbrokers).
Donnison, D. and Soto, P. (1980), *The Good City: A Study of Urban Development Policy in Britain*, London: Heinemann.
Gamble, A. (1981), *Britain in Decline: Political Strategy and the British State*, London: Macmillan.
Glasmeier, A.K., Markusen, A., and Hall, P. (1983), 'Defining High Technology Industries', *Working Paper* 407, University of California, Institute of Urban and Regional Development, Berkeley.
Gudgin, G., Moore, B., and Rhodes, J. (1982), 'Employment Problems in the Cities and Regions of the UK: Prospects for the 1980s', *Cambridge Economic Policy Review*, 8 (2).
Hall, P., (1985), 'The Geography of the Fifth Kondratieff', in P. Hall and A. Markusen (eds.), *Silicon Landscapes*, London: Allen and Unwin, 1–19.
Jones, Lang, and Wooton (Chartered Surveyors), (1983), 'The Decentralisation of Offices from Central London', technical paper.
JURUE (1985), 'Assessment of the Employment Effects of Economic Development Projects Funded under the Urban Programme', *Draft Final Report to the Department of the Environment*, Joint Unit for Research on the Urban Environment.
Knight, Frank, and Rutley (1983), *Office Developments in the Western Corridor*, London.
Langridge, R. (1983), 'Defining High Technology Industry', *Working Paper*, Department of Economics, University of Reading.
Law, C.M. (1983), 'The Defence Sector in British Regional Development', *Geoforum*, 14, 169–84.
Lovering, J. (1985) 'Regional Intervention, Defence Industries, and the Structuring of Space in Britain: The case of Bristol and South Wales', *Environment and Planning D: Society and Space*, 3, 85–107.
—— (1986), 'Defence Expenditure and the Regions—The Case of Bristol', *Built Environment*, 11, 193–206.
Malecki, E.J. (1984), 'Military Spending and the US Defense Industry: Regional Patterns of Military Contracts and Subcontracts', *Environment and Planning C*, 2, 31–44.
Markusen, A. (1983), 'High-tech Jobs, Markets and Economic Development Prospects: Evidence from California', *Built Environment*, 9, 18–28; reprinted in P. Hall and A. Markusen (eds.), *Silicon Landscapes*, London: Allen and Unwin, 35–48.

—— (1984a), 'Defense Spending: A Successful Industrial Policy', *Working Paper* 424, University of California, Institute of Urban and Regional Development, Berkeley.

—— (1984b), 'Defense Spending and the Geography of High Tech Industries', *Working Paper* 423, University of California, Institute of Urban and Regional Development, Berkeley.

Marquand, J. (1979), 'The Service Sector and Regional Policy in the United Kingdom', *Research Series* 29, Centre For Environmental Studies, London.

Massey, D. and Miles, N. (1983), 'Sex, Grouse and Happy Workers', *New Society*, 22 Apr. 12–13.

Morgan, K. and Sayer, A. (1984), 'A "Modern" Industry in a "Mature" Region—The Re-making of Management–Labour Relations', *Working Paper* 39, Urban and Regional Studies, University of Sussex.

Rajan, A. (1984), *New Technology and Employment in Insurance, Banking and Building Societies*, Farnborough: Gower.

Redfern, P. (1982), 'Profile of our Cities', *Population Trends*, XX, HMSO, 21–32.

Rees, R.D. and Miall, R.H.C. (1981), 'The Effect of Regional Policy on Manufacturing Investment and Capital Stock within the UK between 1959 and 1978', *Regional Studies*, 15, 413–24.

Saxenian, A. (1980), 'Silicon Chips and Spatial Structure: The Industrial Basis of Urbanisation in Santa Clara County, California', Master's Thesis, University of California, Berkeley.

—— (1983), 'The Genesis of Silicon Valley', *Built Environment*, 9, 7–17; reprinted in P. Hall and A. Markusen (eds.), *Silicon Landscapes*, London: Allen and Unwin, 20–34.

Short, J. (1981), 'Defence Spending in the UK Regions', *Regional Studies*, 15, 101–10.

Soete, L. and Dosi, G. (1983), *Technology and Employment in the Electronics Industry*, London: Frances Pinter.

Stanback, T.M. (1979), *Understanding the Service Economy*, Baltimore and London: Johns Hopkins University Press.

Todd, D. and Simpson, J. (1985) 'Aerospace, the State and the Regions: A Canadian Perspective', *Political Geography Quarterly*, 4, 111–30.

Townsend, P. (1984), 'Inequalities in Health in the City of Bristol', Department of Social Administration, University of Bristol.

3

London: Employment Problems and Prospects

*Nick Buck, Ian Gordon, and Ken Young**

It is often remarked that the term 'inner city problem', with its connotation of a deprived urban core ringed by areas of relative affluence, misrepresents the spatial patterning of social and economic deprivation in the British conurbations and consequently encourages a misdirection of public policy. In London's case, however, there is some truth in the picture of concentrations of poverty and competitive disadvantage at the heart of the metropolis. Looked at in the round, the London region provides the paradigm case of the 'inner city problem'—one, moreover, which market forces and public policy have combined to bring about. The basic distinction between 'inner' and 'outer' London provides a convenient prism through which to view social and economic problems. However, as will be seen, in many senses Greater London has itself become the 'inner city' of a larger metropolitan region with which its links have gradually loosened.

Our starting-point in this chapter is the long-running process of decentralization of population from inner to outer London and beyond, since it is those left behind by this process who constitute the present inner city population. We argue that the population loss which did so much to alarm policy-makers in the 1970s can be understood largely as a response to the spatial patterning of housing opportunities in the London region rather than as a flight from the city. We then review Greater London's employment decline which, while stemming from a number of independent sources, similarly helped to define inner London as being, in common with the inner areas of the other conurbations, an unattractive location for investment. Neither population decentralization nor employment decline need in themselves imply adverse welfare consequences: during the 1960s and the first half of the 1970s London experienced both without suffering any substantial deterioration in labour market conditions. However, during the current recession, Londoners have experienced unprecedentedly high rates of unemployment, particularly in the inner areas. In the third section of the chapter, therefore,

* This chapter is based on research undertaken jointly with John Ermisch and Liz Mills of the Policy Studies Institute. Our main report, published as *The London Unemployment Problem* (Buck *et al.* 1986), includes a fuller analysis of the factors underlying the employment and labour market developments summarized here, together with an examination of planning policies and local authority economic initiatives which are not dealt with in this chapter.

we turn to an examination of the links between employment change, population characteristics, and unemployment. Finally, we draw together these strands in order to sketch out a view of the nature of London's employment problem before concluding with an indication of the main lines of policy which our research suggests.

'Inner' and 'Outer' City

London's social and economic dominance had given it a special role as the engine of change in the British economy for most of the last four centuries, the rise of the northern industrial cities notwithstanding. As a result of this dominance London drew to itself a disproportionate share of the nation's human and financial resources. In the late nineteenth and early twentieth centuries the combination of migration to London from without and the outward flow of Londoners in search of a cheaper, more spacious, and safer environment produced such rapid suburbanization as to evolve the fear (by no means new) that London would eventually consume the whole of southern England.

Between the wars the growth of outer London's population was accommodated by private house-building, and the great surge in owner-occupation began there. Unlike in the nothern conurbations, the public sector played a relatively minor role in housing development, and peripheral housing estates were not a marked feature of London's growth. Those 'cottage estates' that were built were socially selective through high rents and were soon engulfed as London's frontier spread outward. After 1928, and increasingly from the mid-1930s, working-class Londoners were instead accommodated *within* the inner area, not in houses but in flats, and predominantly in the impoverished 'crescent' to the east, north-east, and south-east of the City of London. This spatial division between the owner-occupied and rented sectors (with the private rental sector being more characteristic of inner west London) became self-sustaining as a political determination to maintain separate territories developed. Thus arose the encirclement of Victorian London and its emergence as the paradigmatic 'inner city'.

The establishment of the Metropolitan Green Belt reproduced this encirclement on a larger scale and set the scene for the eventual emergence of Greater London as the 'inner city' of the larger region. While the first aim of post-war policy, the *containment* of London, was thereby achieved, the second, *constraint* outside London, was partially abandoned in the face of unexpected population pressures, the want of an effective apparatus of regional planning, and a latter-day view that the South East had a vital role to play in achieving a higher rate of growth in the national economy.

Post-war policy aimed to reduce Greater London's population, largely by means of planned dispersal to the New (and, later, Expanded) Towns. The mid-1970s saw a sharp reversal of dispersal policies, which were blamed in

Table 3.1. *Net Migration by National, Regional, and Local Stream: London 1966–1971* (males 15+)

Area	Migration stream			
	National (000s)	Regional (000s)	Local (000s)	Total (000s)
Inner London	−0.9	−56.1	−79.9	−136.9
Outer London	−33.3	−68.4	+57.5	−44.2
Greater London	−34.2	−124.5	−22.4	−181.1
OMA	+9.7	+39.4	+14.1	+63.2

Source: Gordon and Vickerman (1982).

some quarters for 'emptying out' London of its more skilled and vigorous workers. The New and Expanded Towns programme, being largely geared to labour demand in those towns, was undoubtedly highly selective, both socially and occupationally, in its operation. However, the programme provided only one, relatively minor, channel for the outflow of population, which was proceeding most rapidly from inner London. This evacuation of the 'inner city' is part of a long-term process of selective migration (and resultant polarization) in which housing factors have been particularly prominent; the analysis of this residential decentralization is of particular importance in understanding London's 'inner city'.

Population decentralization

Greater London's population has been falling since 1939, when the peak population of 8.6 million was achieved, but the rate of loss accelerated markedly in the mid-1960s. Since 1966 the population has fallen by 1.3 million, entirely as a result of the excess of outward movements of residents over in-migration. This outward drift, largely of working-age individuals and their families, has been particularly marked from Inner London.[1] It results from the balance of inward and outward moves in each of three streams of migrants: long-distance movers (the *national* stream) who are moving to new employment and who, in the London case, are mostly moving to or from areas beyond the outer metropolitan area (OMA); a much larger group of short-distance movers (the *local* stream) who are essentially changing houses, typically in search of additional space, and who, in the London case, are mostly moving within the boundaries of Greater London; and a *regional* stream of individuals changing their residence, though not normally their job, of whom the majority in Greater London would be moving out to the OMA. As Table 3.1 shows, it is this regional stream which has been responsible for most of the population loss from *Greater* London, though most of *Inner* London's loss is attributable to the local stream of housing movers.

National stream moves are the only ones to be significantly influenced by employment considerations, and they account for only a small part of London's net population loss. Indeed, in Inner London, where large numbers of young immigrants are attacted from outside the region, inflows and outflows in this stream are more or less in balance. In any case, poor employment opportunities in the inner areas play only a limited role in generating outward movement. Survey evidence suggests that environmental concerns—for example, about violence, vandalism, or (particularly) 'immigrants'—were more important in promoting outward movement. Nevertheless, the majority of outward movers from Inner London have been essentially local movers seeking larger accommodation or opportunities for owner-occupation (Gordon *et al*, 1983). The outward drift from Inner London originates in housing rather than employment factors and essentially reflects the spatial distribution of housing in the required size, price, and tenure categories for those who are able to make such moves. These factors lead propective owner-occupiers from private rented accommodation in inner areas to seek houses in the outer boroughs. For those in the council rented sector, such outward moves are less likely (since most of the stock is in inner areas) and also much harder to achieve. The outward movers are thus particularly influenced by where new building is increasing the stock of housing for owner-occupation.

Regional stream movers, on the other hand, tend to be existing owner-occupiers seeking a combination of better housing *and* a more congenial environment outside of London. Many fewer people are involved in these medium-distance moves but they have accounted for most of the net outflow from Greater London as a whole because the balance of the flows is very one-sided, with few such moves either into or within London itself. Relative house prices are one factor in these moves, since prices for a given type of housing still tend to fall with distance from Central London. However, environmental preferences are also important as is, again, the simple question of where new housing opportunities are occurring. For both local and regional moves, therefore, there is an important link between land availability and population decentralization via private house-construction rates as well as via relative house prices.

Both the major migration streams out of Inner London and from Greater London as a whole thus appear quite unrelated to labour market factors, such as the decentralization of employment which has been proceeding at the same time. Survey evidence suggests that few outward movers within the London region even took into account the possible availability of local jobs in deciding whether and where to move. The immediate effect of an outward move for most of those involved is extended travel to work, and few household heads change their work-place at the same time as moving out, though more wives may give up or change jobs at this time. However, outward movers are likely in subsequent years to change their work-place to somewhere nearer their new home as they become aware of employment opportunities in the outer areas. It seems clear, therefore, that outward moves undertaken essentially for housing or environmental reasons have had a major effect in dispersing labour supply, at least for those groups who are able to make such moves.

In fact, outward migration is highly selective in terms of both life-cycle position and economic status. Young families are particularly likely to make outward moves, both because they are seeking extra space within a fixed budget and because at the family-forming stage residential preferences often switch in favour of a more suburban environment. Conversely, inward movers tend to be young and childless. Outward movement is also selective in being largely confined to those who can afford owner-occupation, particularly in the case of moves right out of Greater London. This selectivity is more clearly seen in relation to earnings levels than conventional measures of social class or 'skill', but it means that the population of inner areas includes increasing proportions of those who can only secure poorly paid jobs and who are effectively trapped in Inner London housing. The effect of inter-area movement is to increase the relative concentration of less skilled workers in London, and particularly in Inner London. At the other end of the class spectrum, there is little sign that gentrification has significantly reduced the net outflow of professional or managerial workers from Inner London as a whole, however obvious the effects in particular districts. For broad areas within the city, social polarization is apparent only in terms of an accumulation of less skilled or lower-paid workers.

Since about 1974 the rate of outward movement from Greater London has slowed down very considerably. The numbers moving from London to other parts of the South East or from Inner to Outer London fell by almost half between the 1971 and 1981 Censuses. Inward moves slowed to a lesser degree, and started to increase again after 1979. Consequently, net losses from Greater London have been cut back to about 35 000 per year since 1980, about one-third of the rate occurring between 1964 and 1974. Both the Inner and Outer boroughs have been affected, although in relation to population size the change has been much more significant for the inner areas. Several factors seem to be involved in these changes. Rising transport costs since the oil price rises of 1970s have made long-distance commuting a less attractive proposition, especially for potential out-migrants in the regional stream. The major factor, however, appears to have been the national recession in the housing market in the years since 1973, reflected in sharply declining rates of private construction and reduced housing opportunities for outward movers.

It is only plausible to expect net out-movement to remain at its present low level so long as this recession in the housing market persists, and even in this situation the pent-up demands of successive cohorts of prospective owner-occupiers are likely to revive outward movement sooner rather than later. Beyond the short term, the safest assumption, whatever the polices pursued

Table 3.2. *Per Annum Employment Changes in London, 1951–1984*

Period	All industries	Manufacturing	Production industries other than manufacturing[a]	Services
1951–61	+17 700	+400	−1 400	+18 700
1961–6	−14 300	−30 700	+4 700	−11 700
1966–71	−54 300	−38 400	−13 200	−2 700
1971–4	−30 800	−49 200	−6 200	+24 600
1974–8	−41 800	−33 100	−5 100	−3 600
1978–81	−34 800	−24 500	−4 500	−5 800
1981–4	−34 500	−39 000	−8 800	+13 300

[a] Other production industries include agriculture, mining, construction, and gas, electricity, and water.

Sources: 1951–71, Censuses of Population; 1971–84, Censuses of Employment.

in London, is that Greater London as a whole and the inner areas in particular will once again experience substantial population losses in their exchanges with outer areas of southern England, although if economic decline continues to be worse in the North these may still be offset by gains of young migrants from other regions.

These trends, as we have said, have been in operation for a century or so and have been particularly marked during private house-building booms. Historically, they were accompanied by the growth of long-distance commuting as Inner London, and the central area in particular, assumed increasing significance as a source of employment. Over the past quarter of a century, however, employment in London has also fallen since, while service employment has increased in line with sectoral changes in the national economy, these gains over the period have been insufficient to offset the considerable fall in manufacturing employment within Greater London. This phenomenon of job loss, which is now characteristic of the metropolis as a whole and is not confined to London's 'inner city' areas, requires a close analysis if meaningful comparisons with other conurbations are to be made.

Job Loss in London

London employment has been subject to a more or less continuous process of change since 1961, although the pace of change has been quite uneven. Table 3.2 shows that employment started to decline around 1961, and accelerated after 1966, but that the rate of job loss has not further increased in the current recession. As the table shows, job losses were concentrated in manufacturing, but it was only in the early 1970s that there were significant gains in services to compensate for the losses in manufacturing. Indeed, from 1974 to

1981, London experienced significant reductions in service industry employment also.

London and the national economy

The persistent decline in London's employment over the past twenty-five years or so has occurred despite an industrial structure which has been consistently biased towards activities with expanding employment nationally. In particular, London should have benefited from the fact that it lacked agriculture or mining and that the manufacturing sector, with declining employment nationally, was increasingly under-represented in London whereas services, which showed consistent gains in employment nationally, were increasingly over-represented in the city. Even within these two sectors, London had relatively more employment in the particular industries with faster rates of national employment growth, or slower decline. If each industry in London had experienced the same rate of employment change as in the country as a whole, there should have been substantial increases in employment, averaging 30 000 per year, during the 1960s and through to the aftermath of the first oil price rises in 1974. Even in the recessions after 1974, only modest losses would have been expected in London, because the national decline was concentrated in manufacturing industries which were by then of more limited importance in London. Over the decade 1974–84 employment in London should actually have increased by the equivalent of 3000 jobs per year if each industry had grown at its national rate.

In fact, however, employment change in London since 1951 has been consistently less favourable than it would have been, had it followed national trends. This was particularly the case between about 1961 and 1978 when differential losses were running at between 50 000 and 100 000 jobs per year. After 1978, however, employment decline in London did not accelerate as it did in most other parts of the country and differential losses were reduced to about 30 000 jobs per year. Since the industrial structure of London employment also protected it from the worst effects of manufacturing job loss, especially in the years 1978–81, the long decline in London's share of national employment actually ceased between 1978 and 1984.

Manufacturing and services contributed more or less equally to differential loss over the period as a whole, despite the fact that manufacturing had the overwhelming majority of actual job losses. In proportionate terms, the rate of differential job loss has tended to be much faster in the manufacturing sector. However, the balance of losses shifted towards services in the 1970s. The worst period for manufacturing was from 1966 to 1974, with a peak loss of 65 000 in 1972–3. The period of highest services losses began later, around 1970, and ended later, around 1977. In the recession after 1978, particularly in the early phase up to about 1981, manufacuturing employment in London suffered no differential losses compared with national change. Because the structural effect was also favourable in those years of deepening recession elsewhere, London manufacturing as a whole performed rather less badly than manufacturing in the country as a whole, probably for the first time since the war. The service sector was not, however, protected to the same degree.

Examination of the pattern of year-to-year changes over the period since 1966 shows that there is a general tendency, not limited to the current recession, for London's differential losses, particularly in manufacturing, to diminish when national employment is contracting. Conversely, in periods of general growth, London has tended to fall further behind the rest of the country. Thus, in the three five-year periods between 1968 and 1983, including two full cycles and one long downswing, 1968–73 saw manufacturing employment in the nation as a whole declining by 5 per cent, compared with 20 per cent in London; in the second period, 1973–8, there was a national decline of 7 per cent, and a decline in London of 17 per cent; while in the years 1978–83, when national decline was 23 per cent, the fall in London was again 20 per cent. Analysis of annual data for the years 1966–84 suggests that the trend was for London manufacturing employment to fall by about 3 per cent per year plus half the national rate of change.

This pattern is not apparent among manufacturing industries elsewhere in the South East and is not simply the result of differences in the industrial composition of employment. It may, however, reflect a concentration in London of more cyclically insensitive *parts* of various industries—for example, headquarters. Another factor may be the extent to which constraints on expansion in London, operating more strongly in periods of general growth rather than decline, are a cause of differential losses in London, particularly for manufacturing.

These employment changes are also markedly different from those experienced in other British conurbations in the 1960s and 1970s, some of which are described elsewhere in this volume. London's employment structure was much more favourable than that of any other conurbation (except the West Midlands in the early 1960s), but its differential losses were very much worse than those for other cities up to 1975 (except for the West Midlands in the later 1960s). The result was that London experienced a heavier loss over this period, particularly in manufacturing, than did any other city. London lost 38 per cent of its manufacturing employment between 1959 and 1975, compared with a loss of 25 per cent for the Manchester conurbation (the next-worst case), a loss of only 34 per cent in Merseyside, and a gain of 4 per cent for Tyneside.

After 1975, however, London no longer had the highest rate of decline, and in the current recession London has had the most *favourable* employment performance in manufacturing of any conurbation. Thus it would seem that the national processes behind the very substantial rate of job loss in the 1960s and 1970s have become of diminishing importance since the onset of recession in the late 1970s. The form of restructuring occurring since then has been less differentially harmful to London. However, it has been almost as harmful in absolute terms, and it makes little difference to London manufacturing workers whether decline is limited to the city or shared with other areas—indeed, the latter decreases their chances of finding employment elsewhere.

Industrial change

A more detailed industrial analysis of employment changes between 1966 and 1981 shows that the tendency for London's share of national employment to decline was very general, with little evidence of systematic variations in trends, at least among manufacturing industries. Only two industries (public passenger transport and leisure services) had more favourable employment changes in London than nationally over this fifteen year period. Other service as well as manufacturing industries showed substantial differential losses. Poor employment performance in absolute terms was almost universal among manufacturing industries up to 1981, and differential performance (in relation to national changes) was consistently bad up to 1976. In most industries, differential change improved somewhat after 1976, as it did for manufacturing in total. The only cases where heavy relative losses were not reduced were electronic engineering, instrument engineering, and furniture. The first of these was one of the few where national growth was occurring.

The pattern of changes shows few continuities. Only food and drink and other manufacturing industry did consistently worse than the average for all manufacturing, and only chemicals, mechanical engineering (by a small margin up to 1976), and vehicles did consistently better, though printing and other metal goods are close to falling into this category. None of the main types of manufacturing in London—such as the traditional crafts, processing activities, or modern light industry—was performing generally better or worse than the others. The industries in which London employment is contracting much faster than employment elsewhere had rather different characteristics, in different periods. Between 1966 and 1971 London's share of employment declined faster in industries with a high national rate of productivity growth, including both electrical and instrument engineering. During this period, when manufacturing decline was concentrated in the inner areas, it appears that inner city plants were tending to lose out in those sectors where restructuring was most intense. In the next decade, however, London's share of employment tended to decrease most in those industries with more favourable employment changes at the national level, and in those whose workers had below-average earnings levels. In this period, when manufacturing employment decline was more widespread through London, the pattern of differential losses across industries suggests that constraints on growth were a more important factor and that low-paying industries may have experienced particular difficulties in recruiting in London.

Within the services sector, substantial growth in employment between 1966 and 1981 occurred in financial and other business services and, up to about 1976, in public services also. These gains were offset by considerable losses in transport and distribution, together with marketed consumer services (in the early part of the period) and public services (in the later part). Over the period as a whole, losses in these service sectors amounted to about half those sustained in manufacturing, and they were of almost equal importance after 1976. The sharpest declines were in sea transport and other personal services (including private domestic service), though both of these were at their most rapid before the mid-1970s. High rates of differential loss were widespread among services, but most pronounced in sea and air transport, goods transport and storage, hotels and catering, and other personal services. London's share of service employment declined most rapidly between about 1971 and 1976, the period when this sector was growing most strongly in the country as a whole. At this time, retail distribution and some of the office services joined the industries with high rates of differential loss in London.

The spatial pattern of change

The spatial pattern of change *within* London has been very uneven. The changes between 1966 and 1981 in manufacturing and in services are shown in Figures 3.1 and 3.2. In both cases it is clear that the greatest losses are concentrated in Inner London, with lesser losses, or even gains, in Outer London. However, the detailed patterns are somewhat different in the two sectors. In manufacturing, all the boroughs in the eastern half of Inner London experienced high rates of job loss. This was partly linked to the closure of the docks, but these boroughs also contained the major concentration of smaller-scale manufacturing in the inner areas. In the western half of Inner London the pattern was less consistent, but there were high losses in Westminster and, from much smaller bases, in Hammersmith and Wandsworth. The main exceptions to the concentration of manufacturing job losses in Inner London were the substantial contractions in the Outer West London boroughs of Brent and Ealing. Losses in these areas accelerated during the 1970s, and the distinction between Inner and Outer London in rates of job loss started to break down. Only the fringes of Greater London to the east, south, and west now have significantly slower rates of decline. For services, the distinction between Inner and Outer London remains pronounced. Losses in the inner areas are associated with the decline of the docks, with areas of rapid population decline, and also with decline in central area employment, particularly in the City of London and in Westminster where changes in both office and specialist consumer services have involved differential employment losses. Gains have been concentrated on the fringes of

Fig. 3.1. Percentage Change in Manufacturing Employment by London Borough, 1966–1981
Source: Censuses of Population.

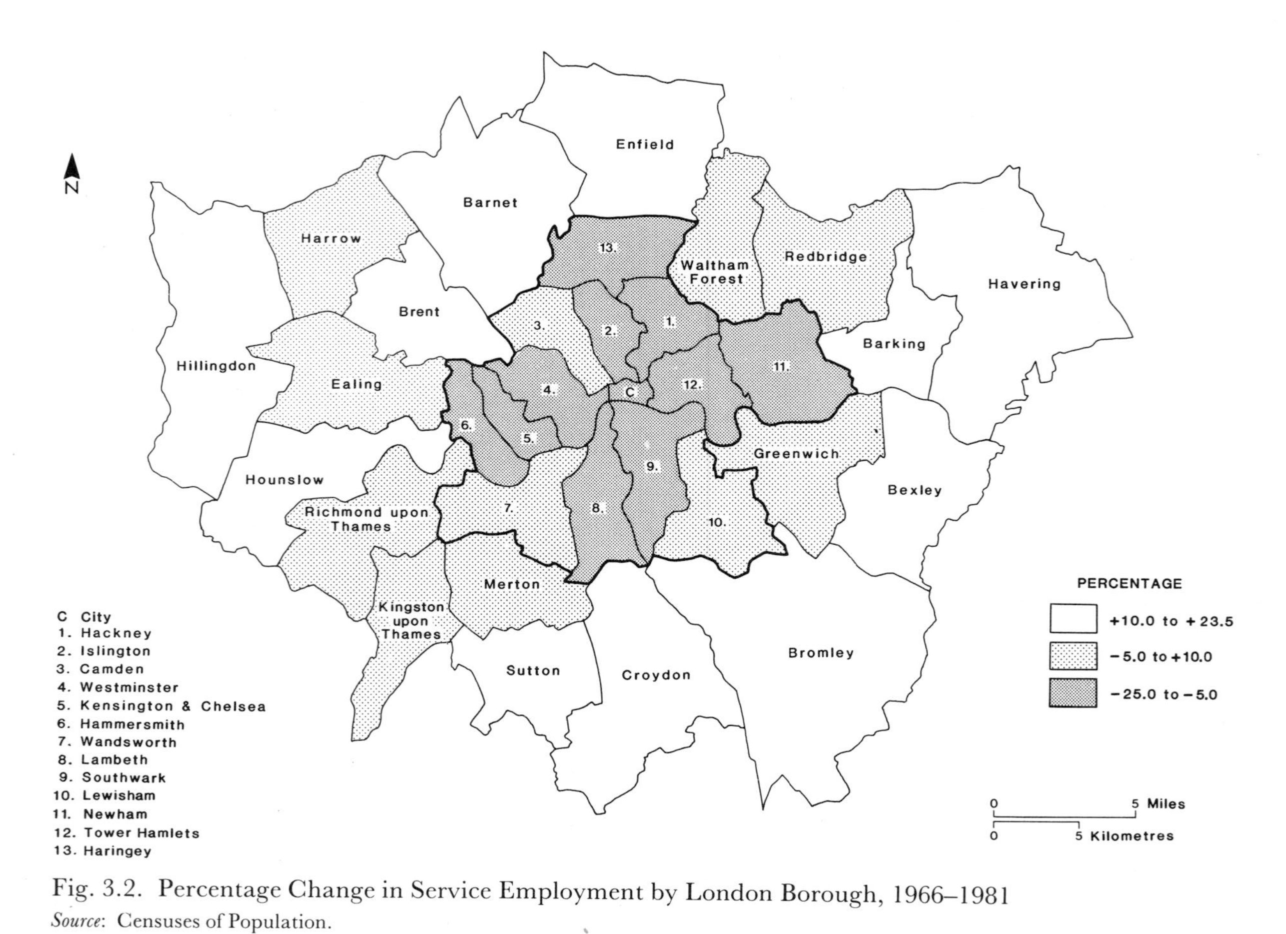

Fig. 3.2. Percentage Change in Service Employment by London Borough, 1966–1981
Source: Censuses of Population.

London, as with manufacturing, though rather more to the north, and less in the south-west. Brent is notable in having a high rate of service growth, in contrast to its high manufacturing losses.

Two study zones within London were selected to provide contrasting examples of economic performance. These zones each included three London boroughs: Inner South-east London, an area developed industrially prior to the First World War, comprised the boroughs of Southwark, Lewisham, and Greenwich; Outer West London, the main area of inter-war industrial growth, comprised the boroughs of Brent, Ealing, and Houslow. The a priori expectation was that change in the south-east zone would reflect the rapid economic decline which most inner areas of large cities have experienced in the last twenty years. Conversely, it was expected that the west zone would exhibit the rather stronger economic performances of the outer areas of conurabations, particularly since it was adjacent to a major area of industrial growth, the Thames Valley and M4 Corridor. This view was reinforced by the fact that the very high industrial rents of the latter area were also shared by the west zone, suggesting a particularly strong demand for industrial premises.

Employment data supports this expectation of contrasting performance for the period up to 1971. In both manufacturing and services, employment change was clearly more favourable in the Outer West than in London as a whole, while the reverse was the case in the Inner South-east. During this period, the Outer West's advantage in terms of relative growth rates appeared to be shifting from manufacturing to services. In the Inner South-east it was always the manufacturing sector which was doing conspicuously badly. After 1971, the contrast breaks down and the two areas display very similar trends. In the manufacturing sector at least, the Outer West was no longer doing better than London as a whole. Neither zone shows the increasing differential losses of service employment during the early to mid-1970s which we noted for Greater London as a whole—these appear to have been concentrated principally in Central London. Nor does either area seem to have enjoyed the substantial protection from the effects of the early part of the current recession (between 1978 and 1981) on manufacturing employment change which other parts of London did. Perhaps as a consequence, service employment in the two areas also showed a marked decline at that time.

By the 1970s, variations in rates of employment change had become more significant *within* the west zone than between the two zones and there was a tendency for the inner parts of this zone to lose jobs faster than the outer parts. This suggests that, if an 'inner city' effect was being felt, it was occurring further out than it had been in the 1960s, in the area which developed in the first three decades of this century as well as in the nineteenth-century industrial areas. A longer-term comparison of rates of manufacturing employment change also supports the view that the area of decline has tended to spread outwards. Up to 1961, manufacturing employment was still growing in the Outer West, at not much below the national rate, although it was declining in the Inner South-east. Since then, however, *both* zones have experienced very substantial declines in manufacturing employment. In the south-east it fell from 128 000 to 47 000 (a 63 per cent loss) between 1961 and 1981, while in the west it fell from 204 000 to 98 000 (a 52 per cent loss). However, the timing of the losses within this period varied, with much the sharpest decline occurring in the south-east before 1971 while in the west the decline of manufacturing has occurred more recently.

Prospects for employment in London

Overall, London does *not* appear to be an intrinsically unviable location for manufacturing industry. Some relatively strong industries nationally have sustained substantial employment in London, and several of the factors which underlay the growth of manufacturing industry in London up to the early 1960s still appear to operate. The first of these is the craft production of consumer goods. However, this is now more or less confined to the luxury fashion-related section of the market (in, for example, clothing and furniture), with production for mass markets being eroded by factory or overseas competition. Secondly, service-related manufacturing continues to be important, both in the case of inputs into services, such as office equipment, and in the case of the information-related printing and publishing industry, though technical change in the latter is likely to lead both to job losses and to the separation of activities, allowing some to move from London. Thirdly, the concentration of control functions in London remains, and indeed has been increasing. In the period from 1965 to 1984, when London lost over half its manufacturing jobs, it increased its share of headquarters of the fifty largest manufacturing companies from twenty-nine to thirty-eight. Continued diversification of companies is likely to reinforce this. However, significant production employment is not necessarily attached to these control functions, and even higher-order functions such as research and development are rather unlikely to be located in London. On the other hand, there has been some evidence of recentralization of multi-plant firms in the recession, and London's predominance as a location of headquarters may allow it to benefit in this process. In general, then, it appears that these three factors are of continued relevance, but to a decreasing proportion of manufacturing activity.

On the other hand, three further factors which once sustained manufacturing in London are no longer of substantial relevance in London and may even operate to its detriment, though they may favour other parts of the London region. The falling population of London, and the selectiveness of out-migration, has substantially reduced the pool of technical and skilled manual labour. This is particularly important in Inner London, where it is reinforced by the decline of manufacturing itself. Some of the Inner London craft industries, such as furniture, used to comprise a highly localized concentration of small firms, with associated workers living nearby. As these concentrations have declined, the labour-supply motivation for locating there has been eroded. On the other hand, the processes of population decentralization have given the OMA an advantage in the supply of skilled, and particularly technical, labour. Secondly, the port has ceased to operate within Greater London, so that industries processing imports have largely been forced out of London (though there are still some riverside factories importing directly). Heathrow airport remains an important transport node, though of more limited significance in affecting the location of manufacturers trading through the airport. Finally, London remains a major consumer market, but falling transport costs, congestion in Inner London, and population decentralization mean both that the benefits of locating close to consumers are less important and that they may be achieved as readily in the OMA (particularly with the completion of the M25) as within London.

London's disadvantages in terms of marginal costs are liable to come to the fore again with the end of recession, when possibilities for reorganizing production and making investments increase, and when cost savings can be less readily achieved through intensification of work. The advantages of the peripheral regions in terms of female labour reserves and assistance from regional policy are less marked than they were in the 1960s. Within the region, too, the cost disadvantages of London have diminished as the rent gradient has flattened out. A London manufacturer would now have to go at least *fifty* miles from London in most directions before finding appreciably lower rents. However, suitable premises are still more likely to be readily available in the outer areas. Moreover, recent survey evidence suggests that the recession has made some manufacturers more conscious of the higher costs in London and that there may be a pent-up demand for movement when investment picks up (Leigh *et al.* 1982). Thus there are reasons for expecting the movement of production from London to continue, though perhaps at a somewhat reduced rate compared with that over the past twenty years.

In the service sector, there are four main processes leading to declining employment in London. Population and business decline in the city act to reduce the demand for many services; the costs and constraints of London locations induce many activities to leave; new technology tends to assist relocation and leads to *in situ* declines, particularly in office employment, though also in distribution; public expenditure constraints have halted the previous rise in public employment and may lead to declines. However, London remains an attractive location for many activities, though to a much greater degree for services than for manufacturing. The high costs are themselves evidence of the attractiveness for some activities. In contrast to manufacturing, the salience of factors favouring a London location for services (the presence of government, the financial centre, the tourism and cultural centre, and the centre of consumption for the most affluent region in the country) has not diminished. There is thus a strong inducement for some sectors to remain in London, particularly financial and business services and those consumer services which serve a market going beyond London (for example, leisure and recreation and hotels and catering). These sectors are the ones which are most likely to experience growth, but this will be limited by costs and spatial constraints, particularly in central London. Their growth within London depends on whether they choose, or can be persuaded, to locate outside the central area.

For most other services the most likely prognosis is decline within London, though not necessarily throughout the London region. The transport and distribution of goods has been subject to many of the same decentralization processes as manufacturing, though more slowly, so that growth was occurring in the 1970s in areas where manufacturing was declining sharply. The process should be accelerated by the completion of the M25 motorway. A decline in population-related services is likely to follow future population declines, and it is also likely, on present policies, that public services will decline at least as fast as population, in contrast to the past, where their relative decline lagged behind. However, it must be stressed that these declines are limited: a precipitate decline of population-related services equivalent to that in manufacturing is entirely implausible.

In summary, manufacturing is likely to continue to decline quite rapidly, though perhaps more slowly than in the last two decades. Its reduction in size means that, in any case, its contribution to job loss will be less than in the past. The same is true of a number of other sectors which have been declining rapidly, such as public transport, sea transport, and public utilities. The decline in population-related services is likely to be broadly in line with past trends, except for the public sector element which is unlikely to make any further contribution to employment growth. Of the 'basic' services, transport and distributive activities are likely to decline more rapidly, while the future of the 'central area' services is perhaps the most uncertain, though past growth appears unlikely to be sustained. On balance, this pattern of change suggests that overall employment trends in London are likely to be broadly similar to those in the past, with neither stabilization nor accelerating decline. The composition of change, however, will be different.

The general patterns of employment change outlined above are likely to have unequal effects in different areas and on different groups within the labour force. In area terms, the past pattern of more rapid decline in Central and Inner London than in Outer London is likely to continue, though the differences should be less great for manufacturing than they have been. In absolute terms, Outer London manufacturing decline will be much greater, given its larger share of employment. On the other hand, the fall in public sector employment is likely to be concentrated in Inner London, a point of considerable importance, given the large share of employment opportunities for Inner London residents it currently provides. The change in private population-related services is also likely to be heavily biased against Inner London. As suggested, Central London office services will probably continue to contract, though a substantial proportion of these jobs are taken by commuters. The specialist consumer services in Central London, which are a greater source of employment for Inner London residents, may continue to grow, though the jobs in these industries are rather low paid and unstable.

The extent to which this continuing shift is likely to have adverse consequences for Londoners is not obvious a priori. It is necessary to consider the nature of London's labour markets and the processes which operate in the face of employment change: their interaction largely shapes the level and distribution of unemployment among Londoners.

Labour Market Factors

Because of the high degree of urbanization in south-eastern England and the strong communications links into London, there is no clearly distinguishable 'London labour market' which can be considered independently of its neighbours. The region, rather, comprises a network of overlapping sub-market areas interacting strongly with each other. The scale of these sub-market areas varies according to the types of job or worker involved. For senior professionals or managers and for many other owner-occupiers, a large proportion of the region might fall within a single travel-to-work area, whereas for the low-paid, council tenants, those with irregular hours, married women in general, and part-timers in particular, effective labour market areas may be highly localized. Nevertheless, for all groups of workers the overlaps are such that developments occurring in a single area are liable to have indirect repercussions throughout the region via adjustments in commuting or residential movement.

Over the past twenty years ar least, the unemployment rate among London residents has remained a little below the national average. In March 1985, for example, it stood at about 12.1 per cent as compared with 13.3 per cent for Great Britain, although in Inner London it was significantly higher, at about 16.5 per cent. These relativities have applied both to men and to women, although, as Table 3.3 suggests, the gap between London and national rates has been widest for non-married women. The increase in London unemployment rates has been very steep during the recession, although, as we have noted, the rate of employment decline in London itself has not accelerated. In the first part of the recession, up to 1982, the rise in unemployment was well below the national average, as it had been in previous downswings, and much less than in the northern regions. Since then, London's position has worsened somewhat. One factor in the recent increases

Table 3.3. *Unemployment Rates in the London Region, 1981*

Category	Inner London	Greater London	OMA	Outer South-East	Great Britain
Persons	10.6	7.8	5.3	6.5	8.8
Men	12.8	9.2	6.1	7.4	10.3
All women	7.7	5.8	4.1	5.1	6.5
Single, widowed, and divorced women	10.5	8.7	7.3	9.0	11.1
Married women	4.3	3.4	2.5	3.0	3.9
16–20 year-olds	20.5	15.7	11.3	13.0	17.3

Note: The unemployment rate as defined here is the number *seeking work* as a percentage of the economically active population resident in the area.

Source: Census of Population.

has been the continuing inflow of young migrants from elsewhere in the country to seek work in London. This inflow itself, however, offers some indication that, despite London's prolonged employment decline, the overall balance between labour supply and demand may not compare unfavourably with other regions.

That view is supported also by the fact that earnings levels for people working in London have been consistently above both the national average and the level in the rest of the South East. Earnings levels in London broadly kept pace with national changes in the decade up to 1978, but since then have grown significantly faster, particularly among manual workers, and faster than in the rest of the South East. By 1984, according to the New Earnings Survey, average weekly earnings in London were 11 per cent above the national average for men manual workers, 17 per cent above for non-manual men and for manual women, and 20 per cent above for non-manual women. The larger differentials for women and for non-manual workers reflect the biases in London's employment structure. Much of the higher earnings may be required to cover higher travel-to-work and housing costs in London, while commuting patterns are such that average earnings of London *residents* are rather lower. Nevertheless, it is significant that during the period of chronic employment decline in London relative earnings levels have not fallen, and that in the recession, when job loss has accelerated elsewhere, relative earnings levels in London have moved upwards.

The general picture which emerges from inspection of the unemployment and earnings data for London, is then, of a regional system of labour markets in which the long-term balance between supply and demand has produced more favourable aggregate outcomes than in most other regions. National recession *has* led to a sharp rise in unemployment in London since 1979, even though rates of job loss did not accelerate as they did elsewhere, but the labour markets appear to have adjusted to the more prolonged process of employment contraction and de-industrialization in the city without continual increases in unemployment.

Effects of employment change

Analysis of year-to-year changes in employment and unemployment since the late 1950s shows that, during the period of generally low unemployment rates experienced up to the early 1970s, the effect on London unemployment of employment changes was very weak, with most of the fluctuations in London being absorbed by migration or commuting changes. In the case of male jobs in the service sector, there was no discernible effect at all on London unemployment. With the much higher levels of unemployment since 1979, reduced mobility has led to employment changes being more directly reflected in unemployment within Greater London, rather than being shifted elsewhere.

Evidence on the effects of redundancies since 1977 shows that involuntary job losses have a particularly strong effect on unemployment. Any new jobs created, even in the manufacturing sector, are quite likely either to be taken up by workers from other areas or to set up a chain of vacancies at the end of which a worker is recruited from outside the area, rather than from the local pool of unemployed. On the other hand, at least when demand is generally depressed, redundant workers are unlikely to see advantages in moving elsewhere and will tend to be well back in the queue for jobs. Thus our regression estimates suggest that redundancies in London manufacturing in recent years have been reflected in more or less equivalent additions to the unemployment register in London.

Allowing for the induced migration which is expected in subsequent years if unemployment is reduced, it appears that, even at the present time, it would require a continuing programme creating 550 jobs *each year* in manufacturing (or a comparable activity) in order to achieve a sustained reduction of 1000 in unemployment in Greater London as a whole. If the target is a smaller and more open area such as Inner London, the effort required would be proportionately greater: for a typical borough it might require around 1100 jobs per year to have that effect on unemployment among local residents. Most of the effect would inevitably get dispersed into surrounding areas.

Effects of population decentralization

The other significant finding to emerge from the analysis of unemployment trends in Greater London is that, if the rates of employment change in London in particular years had been the same as in the country as a whole and if national unemployment had not changed, unemployment rates in London would have tended to *fall* by about one percentage point per year. In other words, alongside—but independently of—the continual fall in job opportunities in the city during the past thirty years, labour supply (both male and female) in Greater London has been shrinking by about 1 per cent per year relative to what would be expected from national changes in population structure and participation rates. Over the long term this shrinkage in supply has more or less balanced with the faster rate of employment decline in Greater London. The crucial factors here are the outward moves of households in search of housing opportunities and their subsequent readjustment of work-place locations in the years following the change of residence. Whether the move is from Inner to Outer London or from Greater London to the OMA and beyond, its initial consequence for most of those involved is extended commuting. Familiarity with the new area will, however, reveal to many a range of alternative employment opportunities closer at hand, to be considered when a change of job is next contemplated or they tire of lengthy commuting.

The effects of outward movement of residents on the numbers seeking employment in the inner areas have to be set against the likely consequences of population loss for local service employment. When this equation is made, it appears that the *short-term* effect of out-movement on the balance between labour supply and demand in the inner areas is likely to be unfavourable but that the *medium-term* effects should be positive and that, in the *long term*, population decentralization has a strongly favourable effect in limiting the rise of unemployment in the inner areas. It is basically a redistributive effect, leaving the numbers unemployed within the London region essentially unaltered. It is, none the less, desirable since there are a range of social, economic, and political arguments for seeking to avoid increasing concentrations of unemployment in the inner city. The most clearly favourable effect would follow from outward moves by workers from lower-income groups, since they both tend to support fewer local service jobs and are less likely to continue commuting to jobs in the centre after a move. However, these are the people who currently find it most difficult to move out of Inner London because of their lack of access to housing opportunities in outer areas and the financial difficulty of commuting from beyond the Green Belt. From a labour market perspective, measures to encourage their decentralization would be particularly desirable, while any factors which stemmed the general outward drift of housing migrants would have undesirable consequences in the long run.

Spatial patterns of unemployment

As in all cities, overall unemployment rates for London hide very substantial variations at a more local level. In 1981, twenty-three wards had rates of between 15 and 19 per cent—twice the average for London as a whole—and these must have risen to between 20 and 30 per cent by 1985. Because of London's size, these concentrations of high unemployment are evident over wider areas than is the case elsewhere. Also, the capital is rather exceptional in the 1980s in that its 'inner city' is still essentially identifiable with the inner areas. All but two of the wards with unemployment of over 15 per cent in 1981 were found within Inner London, and the exceptions were only just beyond its borders, in the inner parts of Brent and Greenwich. A similar pattern could be observed among the larger number of wards with unemployment of over 12 per cent: 85 per cent of these were in Inner London. Very few of these wards were in peripheral council estates as in other cities—in part because inter-war and post-war public housing development in London produced few peripheral council estates.

As Figures 3.3 and 3.4 indicate, the concentration of unemployment in Inner London in 1981 was similar to that which could be observed in times of much lower general unemployment, although it is notable that the worst areas now tend to be to the east rather than the west. Increases in unemployment during the 1970s tended to be proportional to the previous level of unemployment in an area. Hence the differential in unemployment rates between inner and outer areas increased sharply. Since 1981 the position of Inner London as a whole has deteriorated further, in relation both to the outer areas and to Great Britain as a whole, partly at least because of continuing redundancies in both services and manufacturing.

One reason for the considerable, and widening, gap in unemployment rates is the fact that Inner London had until quite recently experienced a conspicuously faster rate of employment loss, particularly in manufacturing, than had the outer areas. However, a detailed examination of the pattern of variation, across boroughs and across wards in London and the OMA, shows that this is not the main explanation. Even in the early 1970s, when employment decline was much faster in Inner London than in the outer areas, this factor appears to have accounted for only one-tenth of the gap in unemployment rates between the Inner and Outer boroughs. It was rather more important, however, in accounting for the difference between the Outer boroughs and the OMA (which had the lowest unemployment rates). At a more local level, ward data from the 1981 Census show that only a very small proportion of the differences is explicable in terms of differing rates of employment change within London.

The dominant pattern is one which reflects the system of residential segregation within the city, rather than spatial variations in labour market conditions, with high rates of unemployment being found in areas of rented accommodation, high density, and concentrations of unskilled manual workers. Tenure emerges as one of the central factors related to unemployment levels, but in London this does not appear to reflect constraints on the ability of council tenants to pursue job opportunities elsewhere (as Hughes and McCormick (1981) have suggested for other parts of the country). Unemployment rates *are* higher in areas with above-average proportions of residents in council accommodation. However, this correlation is most

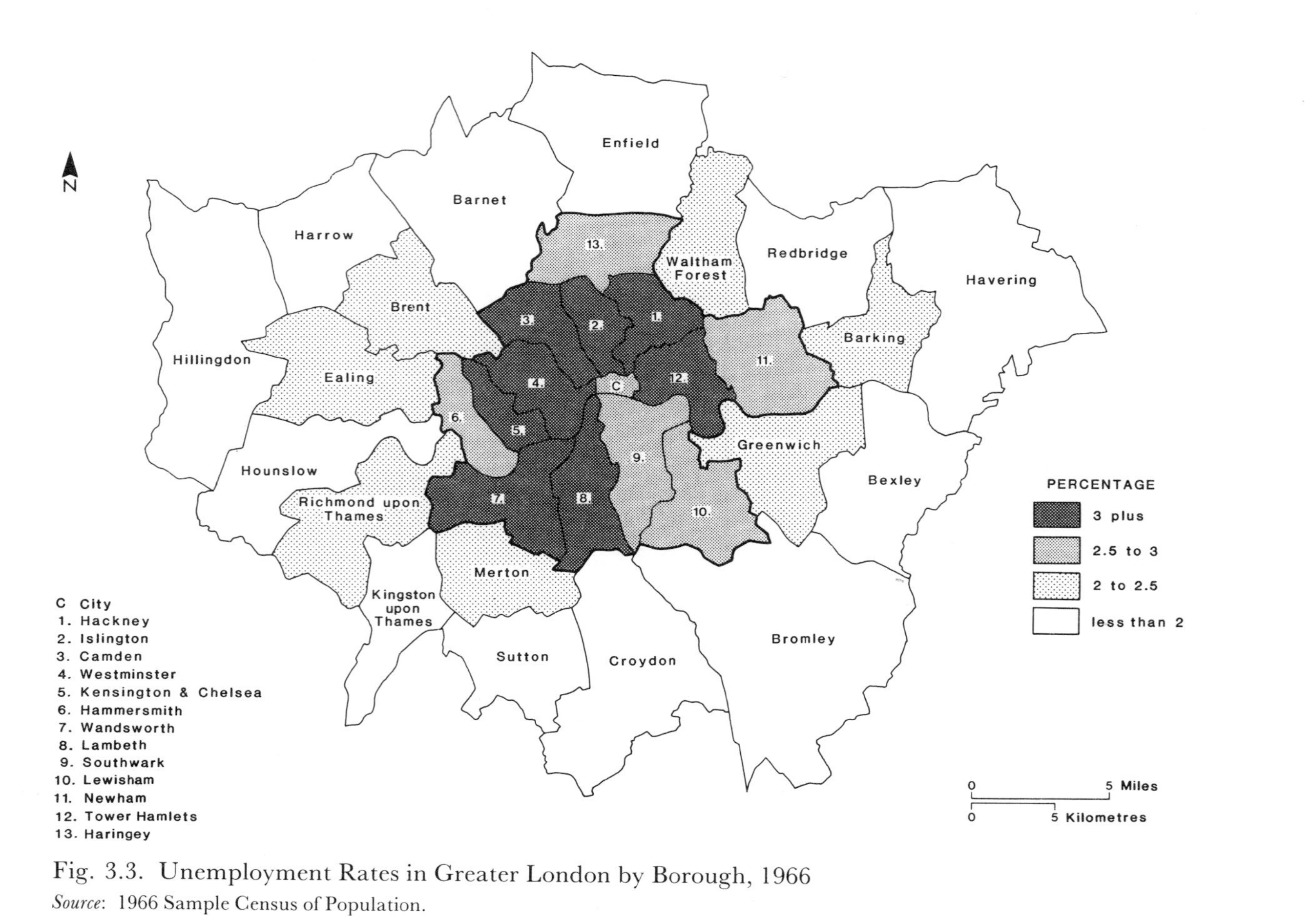

Fig. 3.3. Unemployment Rates in Greater London by Borough, 1966
Source: 1966 Sample Census of Population.

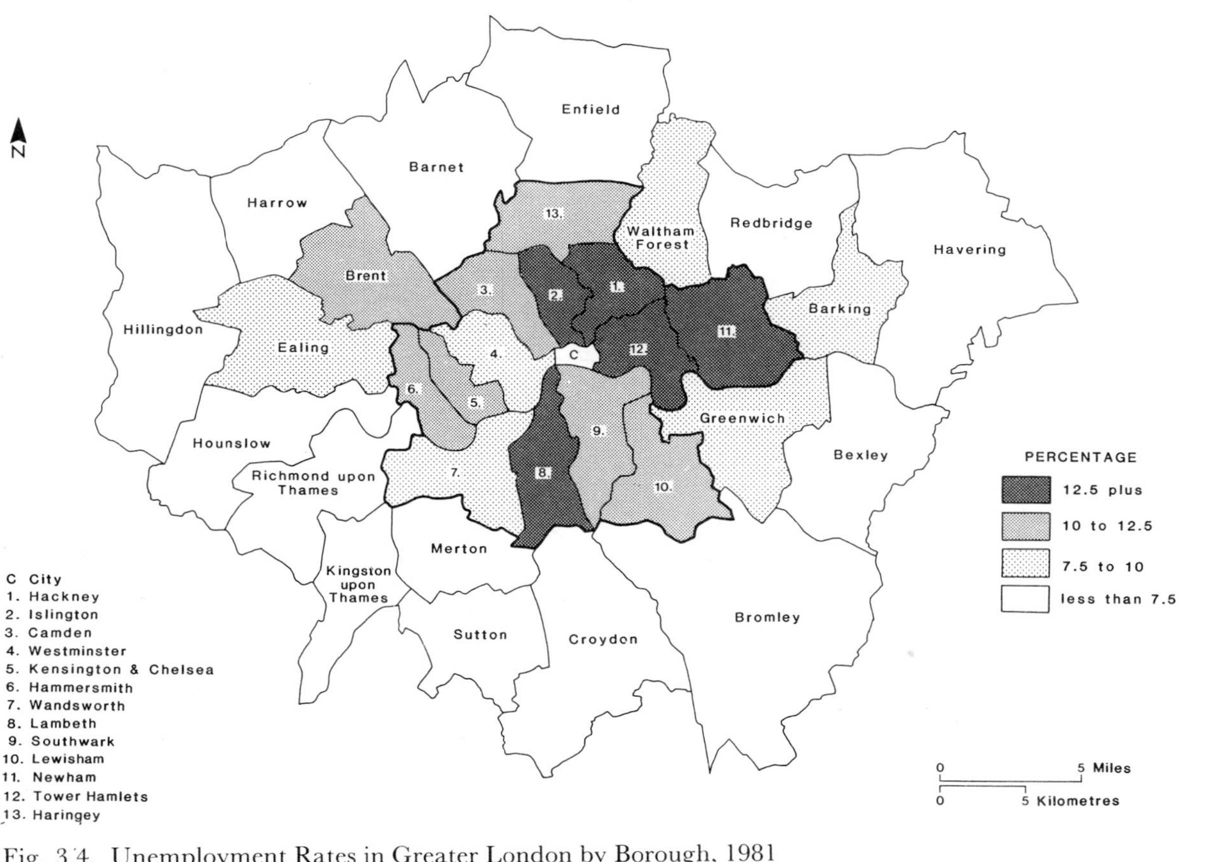

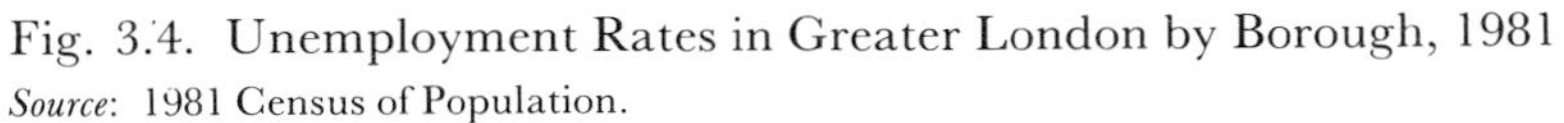
Fig. 3.4. Unemployment Rates in Greater London by Borough, 1981

Source: 1981 Census of Population.

evident at the small area level (that is, wards), reflecting the fine grain of residential segregation, rather than as between broad sub-labour-markets, as would be expected if the problem were one of constraints on mobility preventing labour supply adjustments. Indeed, the connection is most striking at the individual level, with those living in local authority housing being much more liable to unemployment than are owner-occupiers, even when other characteristics such as age, socio-economic group, and industry of employment are controlled for.

Despite a more explicit analysis of the effects of employment change (at the level both of the borough and the travel-to-work area) and sources of labour supply change, our conclusions thus far are consistent with Metcalf and Richardson's (1976) contention that 'Inner city problems occur because individuals who suffer labour market disadvantages live disproportionately in the inner city, because that is where the largest stock of cheap housing (essentially public housing) is found' (Metcalf and Richardson, 1976, 218). Our analyses of ward data from 1981 Census suggest an even wider range of disadvantaging characteristics than Metcalf and Richardson's borough-level study, which emphasized associations with being unskilled, unmarried, or having a large family. For men, and also for unmarried women, we find the most important characteristics related to the incidence of unemployment to be renting from a local authority or housing association, and coming from Afro-Caribbean origins. For married females the most important characteristic is originating from the Indian subcontinent. Race thus appears to be at least as important as class or household structure in affecting individuals' chances of being unemployed.

At the aggregate level, the labour market adjustment processes, of commuting, migration, and occupational change, appear to have operated rather effectively in response to local or subregional variations in rates of employment change. However, they have done nothing to mitigate major sources of inequality in the distribution of unemployment between population groups. The effects of these on unemployment rates have been magnified during the recession, leading to an exceptionally high incidence of unemployment among the young, the less-skilled, and the black population, and across much of Inner London. At the area level these have been reflected in larger increases in unemployment in areas with more black or Asian residents and in those areas with fewer owner-occupiers, as well as in areas where employment has declined much more rapidly than population. In general, high levels of mobility and 'efficient' labour market operations do not help these groups, because selection criteria work against them in competition for jobs with other Londoners and migrants from elsewhere.

Population decentralization provides a particular case of this process. Though stimulated by housing market factors, the outward movement of residents has in the long run served to contain unemployment levels in the labour markets of inner areas which would otherwise have risen substantially as their employment contracted over the past twenty years. However, low income groups (including black workers and the unskilled) have largely been left behind in the general outward movement, both because of constraints on the housing opportunities available to them and because of the cost of commuting from beyond the Green Belt. As employment has declined, competition at the bottom end of the labour market has thus tended to increase, worsening the position, particularly, of unqualified workers in the inner areas. Both housing and planning policies have had their main impact on those without the purchasing power to move freely in the housing market of the wider region and have had an unfortunate effect in enhancing concentrations of unemployment.

Employment stability

One characteristic of London labour markets is their fluidity, which offers some particular advantages to both employers and workers. High turnover rates appear both to shorten expected durations of unemployment among young workers moving between firms and to facilitate upward mobility. Correspondingly, for small firms and those with unstable demands for labour the recruitment problem is eased in a large labour market with many similar firms, and adjustments in numbers employed can be rapidly accomplished. The other side of the coin, however, is increased probability of unemployment for those in unstable jobs and reduced chances of acquiring the qualifications or experience required for a more stable job. These are traditional characteristics of the inner city labour market, as Stedman Jones' (1971) account of casual labour markets in Victorian London underlines. During the present century the London labour market had been partly stabilized by the growth of larger manufacturing concerns, increased employment in public services, and the decasualization of the docks. The collapse of employment both in manufacturing and in the traditional transport sectors has threatened this stability, however, while the growth of employment in large white-collar organizations has provided few opportunities for the unqualified. In the 1970s, only increasing employment in public services served to counterbalance the growing share of jobs in the l[illegible]able private service sectors. This has been particularly the case in [illegible]ner South-east study zone, where the public sector is a particularly important source of employment and had shown substantial growth. Everywhere, however, the effects of rate-support grant withdrawal, rate-capping, and health service cuts have started to undermine this source of stability in employment.

Towards Policies for the London Employment Problem

The London employment problem in the 1980s actually involves the interaction of three sets of problems, centred on the issues of job loss in London in the context of national recession, of the long-term decentralization of population and employment in the region, and of the distribution of employment opportunities within the London work-force. All of these processes have worked against the interests of low-status groups and inner city residents.

Employment decline

The process of employment decline in London is a chronic one which has been running for around a quarter of a century, and not an acute crisis induced by the recession of the late 1970s and 1980s. The national recession has for the time being almost halted the processes of decentralization and restructuring of employment out of London. At the same time, the direct impact of the recession itself on employment in London has been blunted as a consequence both of London's particular employment structure and of the degree of protection being afforded to plants linked to London headquarters. As a consequence, the rate of job loss in London has not accelerated as it has elsewhere. London's share of national employment has more or less stabilized, and the rate of redundancies (especially in the early part of the recession) has been well below the national average.

The basis of this employment decline in London was not, however, primarily structural weakness or lack of capacity to compete. Much of London's large employment loss in manufacturing was not so much a consequence of a high death rate of plants and firms as of a failure to attract new investment and growth within indigenous plants, combined with movement of production out of London, particularly by multi-plant firms. In the 1960s and 1970s, such firms were particularly able to take advantage of lower costs and greater space availability in some areas outside London, while the factors which had made London an attractive industrial location became of declining salience, or came to apply with equal force to a much wider area of southern England. That London lost out in the spatial allocation of new investment was essentially a consequence of market forces. The impact of policy, though its intention was to divert growth from London, was relatively small.

While employment trends in London have thus been quite dissimilar to those in other parts of the country, including the rest of southern England, changes in unemployment have tended to follow national trends rather closely, if with somewhat smaller fluctuations. The long-term decline in London employment—against the national trend—and the even sharper decline in inner areas have been accomplished without a significant upward shift in London unemployment. In contrast, the national recession since 1979, in which London has merely followed national trends and in which the inner areas have not done particularly badly, has brought unprecedentedly high levels of unemployment, particularly in Inner London. Clearly, as far as the labour market is concerned, London is far from being an island, and the greater part of recent increases in its levels of unemployment reflects shifting patterns of labour migration induced by the more severe effects of recession elsewhere in the country. It follows that there is no solution to London unemployment independent of a solution to the national problem. Only a national economic recovery can produce a substantial reduction in London unemployment: no policy action at the London level could conceivably have that effect.

London industry remains an important target for sectoral policies to save some of the valuable productive capacity which is being destroyed in the current recession and to guide structural change. The second important front for action in London is investment in the infrastructure of the city. A boost to construction activities would serve to generate employment specifically for male manual workers who are the group worst hit by the recession.

Job creation in London is, however, only likely to have a limited impact on unemployment among Londoners since the labour market is so open—and this is particularly true if the target is some smaller (and still more open) area such as the high unemployment concentrations of Inner London, or still more a specific borough's population. Job creation policies at the local level are unlikely to do much for the employment prospects of local residents unless they can be targeted at their specific characteristics. On the other hand, in the current situation a very strong connection is evident between the local rate of redundancies and additions to unemployment in the area, implying that effective policies of job preservation, particularly for manual jobs, could have a much more significant impact on local unemployment.

The character, as well as the number, of jobs available in London is also an important determinant of the level of unemployment. Those most at risk of unemployment include workers in small-scale private services and other activities characterized by instability of employment conditions and high rates of turnover. The decline of manufacturing in London, coupled with recent reductions in public service employment, have increased the dependence of unqualified workers on this sector of employment, which has included almost the only elements of growth during the current recession. This trend threatens a re-casualization of the London labour market and the perpetuation of high levels of unemployment among an unskilled work-force. Employment policies in London need to pay attention both to the maintenance of employment in more stable primary sectors, of which public services are now particularly important, and to the stabilization of employment conditions in other service activities.

Decentralization

The long-term trend of employment decline in London is part of a more general process of decentralization in the region which has affected population as much as employment. A dominant factor in this process has been the limited space within the more fully developed areas of the conurbation for activities or households whose demands for space have increased with rising productivity and real incomes. So long as these basic conditions obtain, continuing decentralization is to be expected for all activities or population groups except those which have a particularly strong demand for centrality. The present recession, which has greatly slowed the outward shift of both people and jobs, marks an interruption to the trend rather than a fundamental change of course. A revival of outward migration from Greater London is to be expected whenever the private housing market recovers, and would have a beneficial effect at least in preventing growing inequalities in unemployment as between the inner and outer parts of the region. Opportunities for outward movement have themselves been very unequal, however, with the less-skilled or lower-paid forming a very small proportion of out-migrants, whether in the New and Expanded Towns scheme or in the private market.

Disadvantage

Spatial variations in unemployment rates within London principally reflect the residential distribution of disadvantaged groups rather than the effects of differing rates of employment decline. Very high rates of unemployment in Inner London are attributable to the combination of high national unemployment with competitive disadvantages in the labour market experienced by groups whose weak position in the housing system also has led to their residential concentration in inner areas. The effects of residential segregation are compounded by the lower educational achievements of school-leavers in deprived inner areas. The effect of the recession has been to magnify existing disparities in the incidence of unemployment, and the areas of predominantly local authority housing on the east side of Inner London now have substantially higher unemployment rates than most of Outer London. Similarly, the increases in unemployment have been much greater among the unskilled, blacks, and youths. For a substantial reduction in the disparity of unemployment rates within London, it would be necessary (failing a general reduction in unemployment) to alter the bases on which employment decisions were made in the labour market, to improve education performance in the inner areas, and/or to reduce the degree of residential segregation in the housing systems of the region.

Directions of policy in London

In terms of rates of unemployment for areas of comparable size, the problems of inner areas in London are clearly less acute than those in some northern or midland cities. Such comparisons do not make the London problem any less pressing but they do indicate that it would be difficult to justify ameliorative policies in London which were to any substantial degree achieved at the expense of employment in the more depressed regions. Both for this reason and because London's unemployment problem is not primarily a problem of lack of growth in its region, we do not believe that extension of conventional regional policy measures, such as locational incentives, to Inner London is an appropriate response to the current situation. The implication is, rather, that policies to relieve concentrations of unemployment in London should either be such as to contribute substantially to the national stock of employment or involve primarily a redistribution of opportunities or unemployment within southern England.

Our analysis suggests three principal lines of policy consistent with these aims: job preservation in London; facilitation of the decentralization of low income groups from inner London; and positive action to eliminate discrimination within the London labour market. These three lines of policy are complementary, rather than being alternatives, and each contributes in a different way to the three broad objectives for urban employment policies identified by Davies *et al.* (1984)—namely: restructuring of the economy: amelioration of the effects of decline; and redistribution of opportunities and benefits.

The point has been made earlier that *job preservation* is potentially more effective as a means of countering rising levels of local unemployment than is job creation, because in circumstances of generally high unemployment the redundant are likely to remain out of work within the local labour market, whereas the filling of new jobs sets in motion a chain of vacancies which are likely to end up with the recruitment of workers from outside the area. Job preservation strategies have the potential at least to keep in existence productive capacity in the form of plant, managerial organization, a trained labour force, an existing product range, and/or a network of market outlets which it would require much greater investment to reproduce in a new enterprise. (A new enterprise of any size would also be much less likely to locate in London.) The requirements to achieve this in particular cases include: the provision of finance to weather a crisis, undertake a process of reorganization, or carry out re-equipment; assistance to relax constraints on expansion or relocation within London; and ensuring that appropriate technological, organizational, or marketing advice is available to the firm. Selective intervention of this type based on sectoral analyses is a function which demands both expertise and a 'critical mass', requiring the continuation, in *some* form, of the Greater London Enterprise Board. Its potential is not simply to ease job loss in London but to contribute to the restructuring requirements of the national economy.

Similar arguments for job preservation can be made in relation to public sector activities, where socially useful services are being provided with the use of resources which cannot (or will not) readily be redeployed to more productive purposes. In London, two important cases in point concern the effects of grant cuts or rate-capping on local authorities and of National Health Service cuts or budget re-allocations on health authorities. The same inner areas which have qualified for special assistance under the Inner Urban Areas Act have generally experienced the largest reductions in central government grants or in their permitted expenditure levels. This inconsistency of stance has been criticized on the grounds that it can effectively prevent the intended 'bending of main programmes' and that these are areas of generally greater need. However, particularly if such cuts lead to redundancies in areas of high unemployment, there is also an argument to be made in terms of the value of job preservation. Jobs saved in public services where no substantial efficiency gains are involved will tend to be 'additional' and free from 'deadweight' in a way which can very rarely be assured when private firms are being assisted.

The second important line of policy adaptation to which our analysis leads is the facilitation (and encouragement) of *outward movement* by those lower-paid workers who at present experience great difficulty in leaving Inner London. The main argument for this approach is that population decentralization leads over time to the dispersal of labour supply and, thus, to the reduction, or containment, of unemployment levels in inner areas. To this extent it is principally a redistributional argument as between the inner and outer parts of the London region. However, the reduction of concentrations of unemployment probably has social effects which are more than merely redistributional. Moreover, as the Lambeth Inner Area Study argued (Shankland *et al.* 1977), the lower-paid deserve as much freedom of choice as those who can afford owner-occupation in deciding whether they wish to remain in Inner London, to move to the suburbs, or to move right outside Greater London. If in the course of time this led to a more balanced social mix across London, it could well have additional benefits in terms of educational achievement in currently disadvantaged areas and in reducing the stigmatization of these areas in the eyes of prospective employers.

At present, there are two principal obstacles to outward movement by low-income households. The first of these results from the quite sharp divide in the housing system between the two major tenure systems, council housing and owner-occupation, in terms of access conditions and the geographical distribution of opportunities. Lack of public housing opportunities in the outer areas, or alternative tenures for those who cannot afford conventional owner-occupation, and the continuing difficulty in arranging transfers between housing authorities combine to trap low-income families in the areas of most rapid employment decline, in Inner London, and particularly in its eastern half. Housing policies in London need to give much greater weight to releasing these constraints and breaking down the geographical and institutional division between 'owner-occupied London' and 'council-rented London'. Clearly, there are, in both Inner and Outer London, vested political interests in preserving the status quo, but overcoming these is quite fundamental to any resolution of the 'inner city problem' in London.

The second is the Metropolitan Green Belt which has had the dual effect of raising house prices in Outer London above the level which most manual or service workers can afford, and of imposing higher travel-to-work costs for those seeking moves right out of London, which are then also beyond the means of the lower-paid. While it was active, the New and Expanded Towns policy offered a loophole for this belt of constraint, although employers' priorities made this also quite socially selective. The aim now would not, however, be to promote linked decentralization by London workers and firms, but only to facilitate population movement as a means to decentralization of labour supply. This is a two-step process which requires, firstly, the availability of housing opportunities within travel-to-work range of existing work-places and, subsequently, the availability of local employment opportunities to which movers may later switch. In the London case, these conditions can apparently only be met by the opening up of wedges of land in the Green Belt to mixed developments of private and public sector housing, primarily for London workers, and industry or other employment-generating activities which would otherwise have located further out in the South East. Again, there would clearly be opposition to such a redirection of policy from those who have benefited from blanket restriction, but, if development in the Green Belt is to make a positive contribution to London's problems, it is important that it should occur in a planned fashion within the context of a strategy for the region rather than via a process of attrition.

The third line of policy innovation which is required to meet the London employment problem is more explicitly redistributive, involving *positive action* to eliminate discrimination against disadvantaged groups in the labour market, including black workers, the unqualified, and, in some contexts, possibly all residents of inner areas. The issue is not simply one of targeting new jobs specifically at workers from these groups or creating protected enclaves of employment outside the main stream of the labour market, but rather one of seeking to 'bend' the operations of the market so that they operate less consistently to their disadvantage. At a practical level this would involve, for instance, seeking to give a greater priority (as the law permits) to black youths, the unqualified, and inner city residents in the allocation of those training opportunities which are more likely to lead to employment rather than deferred unemployment. Also important would be a more rigorous application of equal opportunities policies to placement practices in labour market agencies.

In the case of both local government and the rest of the public sector (where there has been generally less concern to 'open up' employment), the most crucial factor will continue to be the number of posts and vacancies to be filled. Equal opportunity policies will prove a chimera if the volume of openings to which they provide access remains insignificant. Thus, to increase public sector employment levels in inner London is the most direct and the most readily available policy option for alleviating the London employment problem in the late 1980s.

The successful pursuit of all or any of these lines of policy intervention would require a much stronger commitment to the improvement of opportunities for Inner London residents, and disadvantaged groups throughout the city, than has been forthcoming in earlier phases of urban policy. In addition, there are a number of institutional prerequisites which will have particular significance after the abolition of the Greater London Council (GLC). Effective intervention to preserve employment requires staffing resources in terms of numbers and specialization, to ensure robust assessments of applications for assistance, which cannot be justified at the borough level. The record of the Greater London Enterprise Board indicates the clear superiority of a broader-based and larger-scale agency in this field. Implementation of a strategy of population decentralization to improve employment opportunities for disadvantaged groups similarly requires planning and housing authorities operating over areas closer in scale to those of housing and labour market processes, and an organizational framework for regional planning across the metropolitan region. The fragmentation of London government will be an even greater handicap in this respect after the demise of the GLC than it has been in the past twenty years. This would be the case also for redistributive policies whose success depends in part on the extent to which single authorities accommodate both groups with particular needs (for example, of improved access to employment) and the resources with which they can be satisfied. Such arguments point to the need for consolidation of the thirty-two London boroughs and the City into a smaller number of all-purpose authorities organized on a radial basis to bring together inner and outer areas and to reflect the main lines of communication and population movement within the city (see Young 1984). If the land-use planning and economic role of the home county councils were also strengthened *vis-à-vis* that of the districts, the prospect would arise for the first time of an institutional pattern adequate to the control and co-ordination of metropolitan development—for an ability to match the scale and range of activity of other actors in the land and labour markets of the region is required in order to achieve more favourable outcomes for disadvantaged residents of inner areas.

Note

1. Inner London has been defined in different ways by various authorities over the past twenty years, but in each case it has included the City and about a dozen of the thirty two London boroughs. Wherever possible in this study we have used the 1981 Census definition which includes the City, Camden, Hackney, Hammersmith, Haringey, Islington, Kensington and Chelsea, Lambeth, Lewisham,

Newham, Southwark, Tower Hamlets, Wandsworth, and Westminster. Outer London comprises the remaining boroughs within Greater London. This definition places Greenwich, the outermost of the boroughs in our Inner South-east study area, in Outer London, although it is in administrative terms an Inner London borough: that is, it falls within the Inner London Education Authority (ILEA) area.

References

Buck, N., Gordon, I., and Young, K, (1986), *The London Employment Problem*, Oxford, Clarendon Press.

Davies, T., Mason, C., and Davies, L. (1984), *Government and Local Labour Market Policy Implementation*, Aldershot: Gower.

Gordon, I.R. and Vickerman, R.W. (1982), 'Opportuity, Preference and Constraint: An Approach to the Analysis of Metropolitan Migration', *Urban Studies*, 19, 247–61.

—— —— Thomas, A., and Lamont, D. (1983), *Opportunities, Preferences and Constraints on Population Movement in the London Region*, Final Report to the Department of the Environment, Urban and Regional Studies Unit, University of Kent at Canterbury.

Hughes, G. and McCormick, B. (1981) 'Do Council Housing Policies Reduce Migration between Regions?'', *Economic Journal*, 91,919–37.

Leigh, R., North, D., Gough, J., and Escatt, K.S. (1982), *Monitoring Manufacturing Employment Change in London 1976–1981: The Implications for Local Economic Policy*, London Industry and Employment Group, Middlesex Polytechnic.

Metcalf, D., and Richardson, R. (1976), 'Unemployment in London', in G. D. N. Worswick (ed.), *The Concept and Measurement of Involuntary Unemployment*, London: Allen and Unwin.

Shankland, G., Willmott, P., and Jordan, D. (1977), *Inner London, Policies for Dispersal and Balance*, London: HMSO.

Stedman Jones, G. (1971), *Outcast London*, Oxford: Oxford University Press.

Young, K. (1984), 'Governing Greater London: The Background to the GLC Abolition and an Alternative Approach', *The London Journal*, 10, 69–79.

4

Urban and Regional Policies and the Economic Development of the Newcastle Metropolitan Region

*John Goddard, Fred Robinson, and Colin Wren**

Introduction

One of the most striking characteristics of the economic geography of the United Kingdom (UK) is the diversity of local economic structures, problems, and policy environments. From the outset, the Economic and Social Research Council (ESRC) Inner Cities Research Programme has been concerned to explore this diversity, to examine the differing responses of particular places to economic change, and to assess the contribution of the various policy measures applied in different areas. Hence, the programme adopted a case-study approach, and research on four conurbations was initially commissioned (Clydeside, West Midlands, Bristol, and London). Subsequently, the Department of the Environment agreed to finance a further case-study, of the Newcastle metropolitan region, which would also contribute to the department's research on the Urban Programme and economic development policies. The addition of the Newcastle case-study helped to increase the range of types of areas studied in the ESRC Programme by providing an example of an English conurbation subject to both urban and regional policy, thus offering an opportunity to explore the aims and impacts of these two policy approaches.

More specifically, the research attempts to assess the comparative effectiveness of regional and urban policy instruments—particularly the provision of direct financial assistance to industry—in the economic development of the metropolitan region, and seeks to identify complementarities and conflicts between policies. It focuses on the degree to which policy serves to strengthen the local economy and provide employment, particularly for economically disadvantaged residents of the inner areas which are designated under urban policy. The ultimate aim is to recommend ways of increasing the effectiveness of policy and improving delivery mechanisms.

The definition of economic development policy is not self-evident, nor can

* The authors gratefully acknowledge the other members of the Newcastle city study research team: David Storey, John Howard, Fiona Waterhouse and Liz Holmes.

it be clear-cut. Virtually all areas of public policy have implications for local economic development: public expenditure generates employment and incomes; macro-economic policy sets the context within which firms operate; and many of the programmes of central and local government indirectly affect economic activity. While not denying the validity or value of this broad-based view of policy, here we largely confine our attention to those policy interventions *primarily* and *explicitly* oriented towards achieving regional or local economic development objectives. In the study area this includes regional policy, public sector provision of factory premises, and local authority economic policy initiatives, such as the provision of sites and premises, financial assistance to industry, and support for small-business agencies.

In general, economic development policy has been subject to little evaluation, despite its increasing adoption and development, especially at the local level. There is no doubt that evaluation is difficult because of the existence of a wide range of policies and measures which frequently change and which may be conflicting or complementary. It becomes hard to isolate individual policies and their effects. Above all, it is difficult to follow through the impacts of policy to the level of the firm, the establishment, and the employees: effects are disparate and often seem elusive. In addition, policy objectives are often unstated, vague, or contradictory, and are themselves subject to change; and, with policies which represent a simple political commitment to 'do something', there is frequently not much commitment to evaluation. However, policy-makers do have to make choices and select policies, and for this reason, if no other, evaluation would seem vitally necessary.

This chapter provides an overview of economic development policy in the study area in relation to its evolution, implementation, and the extent to which it addresses the area's problems. We consider the extent to which policies serve to promote and fulfil 'industrial' objectives, such as restructuring and increased competitiveness, and serve to generate employment especially for deprived residents of the area. The concern with labour market impacts bears on the issue of spatial targeting: Do policies targeted on 'inner city' firms benefit 'inner city' residents? What are the spatial impacts of regional policies which operate throughout the study area without an inner city dimension? How might policy be more effectively delivered and better tailored to meet the needs of the area?

In attempting to answer such questions we have had to assemble data from a wide variety of public sources and from our own surveys. Basic information on which firms and areas are assisted by central and local government agencies, the amount and form of this assistance, and the expenditure on other economic-development-related initiatives such as factory building, advisory services, and training are not readily available and had to be extracted manually from original records. The assessment of the labour market implications of financial assistance has involved questionnaire surveys of employers and employees in assisted firms. The research has therefore sought to identify the economic development problems of the area, the agencies that are seeking to address these problems, their objectives, the policy instruments that they have adopted, and the total financial resources deployed; finally, the study has assessed the extent to which this pattern of deployment has succeeded in addressing the area's problems. Where the research has revealed a mismatch between problems and policy outcomes, we consider the explanation in terms of limitations in resources, powers, and procedures for policy implementation—basically, the capacity of local institutions to effectively address the twin problems of restructuring of the urban economy and the provision of employment for its most deprived residents.

The Newcastle Metropolitan Region

The study area has been defined with respect to labour market relationships and the application of regional policy. It comprises the 'functional region' (Coombes *et al.* 1980, 1981) centred on Newcastle upon Tyne, and includes the conurbation's commuting hinterland and covers those areas consistently subject to regional policy over the period 1972–84. The functional region concept has the advantage of defining an area which recognizes the interdependencies of an urban core and the surrounding settlements, providing a framework for examining a 'local economy'. It recognizes, in particular, the interrelationships of the local labour market and also provides a diversity of economic structures and policy environments for comparative analysis.

This functional region does not constitute a single administrative entity: it is comprised of eleven local authority districts and incorporates the whole of the metropolitan county of Tyne and Wear and parts of Northumberland and Durham (Figure 4.1). The main service centres within the Tyne and Wear conurbation are Newcastle itself, Sunderland, Gateshead, North and South Shields, and Washington New Town. Outside Tyne and Wear County there are extensive rural areas, but also a number of subordinate centres, such as Durham to the south and Morpeth to the north and the local authority-sponsored new town of Cramlington in Northumberland. The south-western part of the study area includes Consett, in Derwentside District, which experienced a major steel closure in 1980, while the eastern part of the area contains nine collieries on the active periphery of the coalfield. The population of the Newcastle metropolitan region is 1.56 million (1981), and Tyne and Wear County, with a population of 1.14 million, accounts for nearly three-quarters of this total.

Many of the area's economic problems are long-standing, and their origins can be traced back to the nature and pattern of industrialization in the nineteenth century. The area has long been identified as a 'distressed', 'depressed', or 'problem' region (or, more euphemistically, 'less prosperous' or 'less favoured') and has a history of regional policy intervention which

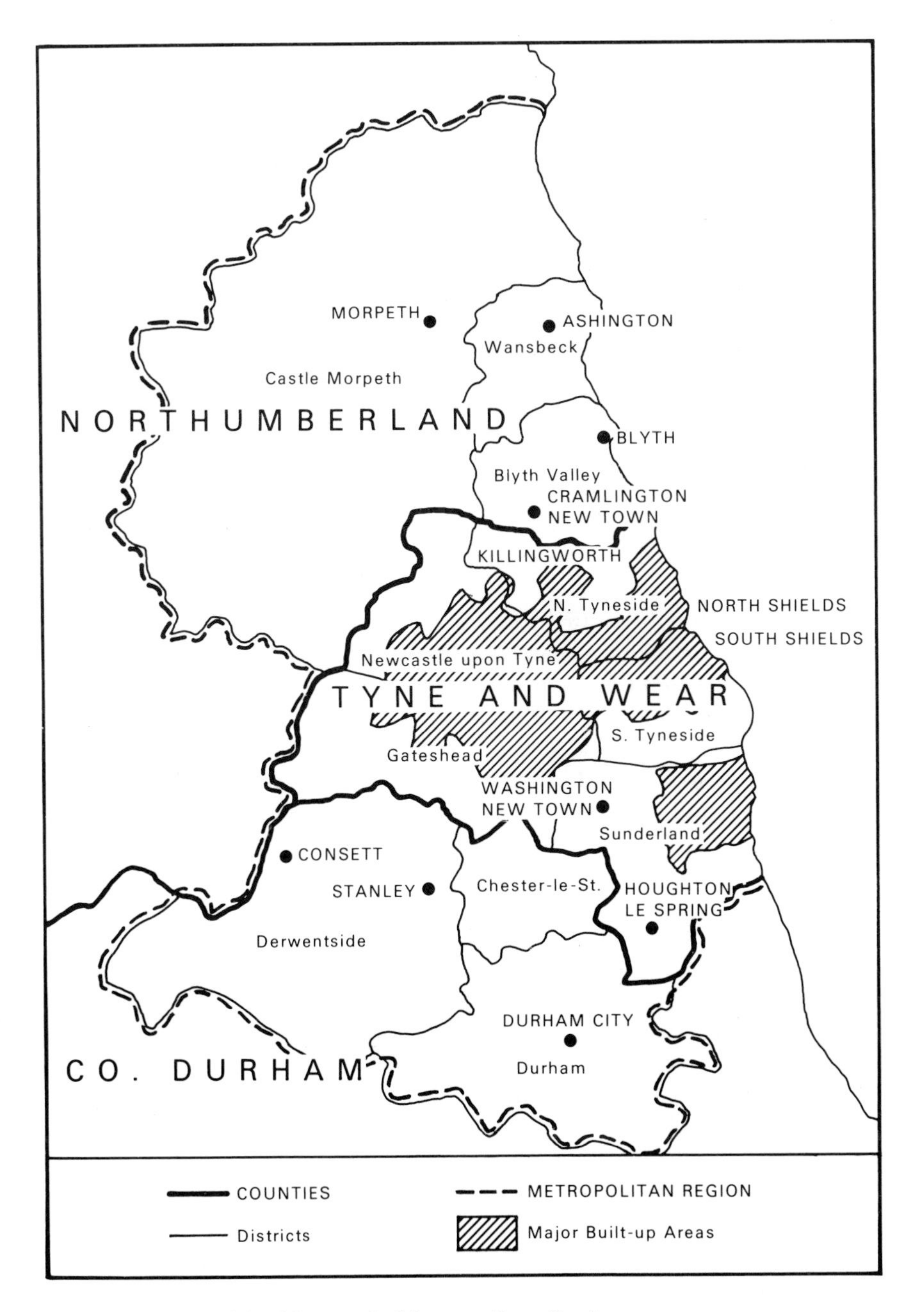

Fig. 4.1. Map of the Newcastle Metropolitan Region

stretches back more than fifty years. It has been described as 'a museum piece of the former glories of industrial Britain' and a 'laboratory for regional development policies' (House 1969, 15). More recently, with the shift of attention towards 'urban' problems, all the districts in Tyne and Wear have been designated for assistance under the 1978 Inner Urban Areas Act. Although the primary focus of our research has been on the period 1974–84, the next section of this chapter considers the antecedents to the area's problems and past responses to them—not least because the long history of industrial decline and the sequence of policy initiatives and related agencies, laid one upon the other, have in many senses also become part of the contemporary problems of the area.

Historical Context: Economic Development and Regional Policy

The basis for the industrialization of the Newcastle metropolitan region was coal-mining, which developed initially along the banks of the rivers Tyne and Wear. Subsequent technological innovation in transport (first the wagonways, later the railways) and in mining itself allowed the substantial expansion of the coalfield throughout the nineteenth century. The second half of the nineteenth century witnessed a massive industrial expansion, with continued growth in coal-mining and very considerable development of manufacturing industry (McCord 1979). By the 1890s, shipyards on the Tyne and Wear accounted for three-quarters of tonnage built in UK yards. The area became an important centre for heavy engineering, notably marine engineering, armaments, and turbines. The Consett steelworks, set up in the 1840s, produced 10 per cent of UK steel by the 1880s. Large-scale economic growth was accompanied by rapid urban development, including mining and shipbuilding communities on Tyneside and Wearside and scores of pit villages across County Durham and south-east Northumberland.

As industrial expansion accelerated, so the industrial structure narrowed: skills, enterprise, and capital were concentrated in the staple industries of coal, shipbuilding, and heavy engineering (Mess 1928, 23). This narrow base proved vulnerable to foreign competition and was heavily dependent on continuing high demand for armaments. Tyneside became a 'huge arsenal and dockyard' during the First World War, but once the post-war boom subsided it was plunged into severe depression, which marked the beginning of a long (and continuing) period of decline. Throughout the inter-war years unemployment was high, especially in the shipbuilding towns and the West Durham pit villages. One response was migration: Jarrow's population fell by 37 per cent between 1921 and 1939, while Hebburn lost 25 per cent of its population and South Shields 21 per cent.

Government intervention to ameliorate regional unemployment began in 1928 with industrial transference schemes to promote increased migration, but the impact was limited, not least because of the shortage of jobs elsewhere and because it was also recognized that selective migration could undermine the viability of areas losing population (Odber 1965, 332–3). The 1934 Special Areas Act marked the first, tentative, efforts towards helping the 'depressed areas' directly, initially through infrastructure projects and then through the development of industrial ('Trading') estates and factories to rent, and, in 1937, the provision of loans and grants to incoming industry. Tyneside, Wearside, and County Durham were designated as 'Special Areas', but the effect of these policies before the War was limited. The first industrial estate was started at Team Valley in Gateshead in 1936, but this and other new estates in the area together employed only about 2500 workers by 1939. It was rearmament and the production imperatives of wartime which restored the economy, again 'expanding the basic industries which had caused so much inter-war unemployment' (Pepler and MacFarlane 1949, 43).

The development of wartime planning and the rise of 'labourism' and 'welfarism', allied to the demands of reconstruction, secured a place for regional policy after the War. Most of the study area was designated as part of the North East Development Area under the 1945 Distribution of Industry Act, which provided for the construction of estates and factories in the Development Areas while restricting factory building elsewhere. After early successes—because of the high demand for factory space—the policy weakened after the balance of payments crisis in 1947 and was then in abeyance from 1951 to 1958. However, by 1952 twenty industrial estates had been built in the study area, providing over 26 000 jobs in 184 factories. Half the workforce in these new factories comprised women, thus sustaining the growth in female economic activity begun during the War.

Throughout the 1950s the area experienced low levels of unemployment (2–4 per cent), owing to strong demand in the staple industries; however, falling demand in shipbuilding and the commencement of pit closures raised unemployment and led to the revival of regional policy at the end of the decade. Initially south-east Tyneside and Wearside were designated as Development Districts under the 1960 Local Employment Act. At first the approach was to designate unemployment 'blackspots' (generally areas with unemployment above 4.5 per cent) but, as conditions deteriorated, coverage was extended to include almost all the study area by 1963. Again, the main thrust of policy was industrial estate development, but financial assistance became more important after the introduction of a standard grant system in 1963.

The period from 1963 to the early 1970s was one of major economic change in the North East, which included a large-scale reduction in mining, active regional policy, service sector growth, and infrastructure renewal. The appointment of Lord Hailsham in 1963 as Minister with special responsibility for the North East resulted in the preparation of a co-ordinated programme of action, led by central government, to diversify and modernize the region, underpinned by 'accelerated investment in public services'—notably roads, factories, and industrial estates, but also including town centre redevelopment, housing, and land reclamation. The 'Hailsham Plan' (HMSO 1963) recommended the establishment of a New Town at Washington (designated 1964) and included most of the study area as part of a regional 'growth zone' within which economic development efforts were to be concentrated. Following Hailsham, regional policy and planning were continued and developed by the 1964–70 Labour Government, with expenditure rising from £80 million in 1966–7 to £259 million in 1968–9 (Manners *et al.* 1972, 51). The whole of the study area (indeed, the whole of the Northern Region) was designated a 'Development Area' under the 1966 Industrial Development Act, although the 'blackspot' approach was reintroduced in 1967 by the designation of 'Special Development Areas' (SDAs) offering a higher level of financial assistance. The five SDAs in the area covered the pit-closure areas of north-west Durham and south-east Northumberland, but Tyneside and Wearside were added in 1971. The traditional instruments of regional policy—factories and financial aid—were used, the latter including not only (increased) grants for investment, but also a wages subsidy (Regional Employment Premium, introduced in 1967, abolished in 1977).

The changing industrial and employment structure of the area between 1961 and 1971 is shown in Tables 4.1 and 4.2. Over this period, coal-mining experienced massive decline, with a reduction of nearly 60 000 jobs. Manufacturing experienced relatively little overall decline (−4.4 per cent), with similar rates of loss of male (−4.1 per cent) and female employment (−5.3 per cent) and small changes in most industries apart from shipbuilding (which lost over 16 000 jobs, a fall of 34.9 per cent). The movement of branch plants into the area, supported by active regional policy, helped to sustain and restructure manufacturing and so compensate for job losses within this sector. However, the major element of change was service sector growth, especially in public sector services, such as health and education—Standard Industrial Classification (SIC) 25—and public administration (SIC 27). It was this which *numerically* offset the losses in coal-mining, although the typically much higher proportion of females in service sector occupations meant that there was, nevertheless, a significant decline in male employment over the decade. In contrast, female employment grew both in absolute terms and as a proportion of total employment: by 1971, 37.5 per cent of employees were women, compared with 31.3 per cent in 1961. Total employment fell by only 3.3 per cent between 1961 and 1971 despite huge losses in mining, but it was not so much regional policy which 'delivered' jobs—rather, it was the expansion of public expenditure on services.

In conclusion, while the formal emphasis of public policy during this period was on the North East Region as a whole, a major part of the impact of industrial change and policy was within the Newcastle metropolitan region as defined for this study. A majority of the region's job losses in min-

Table 4.1. *Employment Change in the Newcastle Metropolitan Region by Industrial Order, 1961–1981*

Industrial order (1968 SIC)	Total employment (000s)				% Change in employment		
	1961[a]	1971	1978	1981	1961–71	1971–8	1978–81
1 Agriculture, forestry, fishing	5.0	3.4	2.7	2.4	−33.2	−19.3	−12.9
2 Mining and quarrying	97.1	38.2	28.6	22.6	−60.6	−25.1	−21.0
Total primary	**102.1**	**41.6**	**31.4**	**25.0**	**−59.3**	**−24.6**	**−20.3**
3 Food, drink, and tobacco	19.1	18.9	14.5	12.1	−0.9	−23.0	−16.5
4 Coal and petroleum products	2.1	1.5	1.6	1.4	−25.5	+3.1	−12.6
5 Chemicals and allied industries	11.9	12.6	10.4	9.2	+5.5	−17.3	−11.2
6 Metal manufacture	13.4	11.4	12.0	5.3	−15.1	+5.5	−55.7
7 Mechanical engineering	39.5	39.6	36.2	26.2	+0.3	−8.6	−27.8
8 Instrument engineering	0.7	2.2	2.6	2.3	+200.3	+16.6	−11.0
9 Electrical engineering	28.3	34.1	27.3	21.7	+20.5	−19.8	−20.6
10 Shipbuilding and marine engineering	47.8	31.1	29.5	23.1	−34.9	−5.3	−21.5
11 Vehicles	3.5	3.4	3.1	2.4	−2.9	−8.4	−21.9
12 Metal goods n.e.s.	6.3	6.5	7.7	5.2	+3.7	+19.2	−33.2
13 Textiles	3.8	6.4	5.7	4.6	+67.1	−9.9	−18.9
14 Leather, leather goods, and fur	1.0	0.9	0.4	0.2	−12.6	−54.1	−39.5
15 Clothing and footwear	16.8	15.7	13.2	8.3	−6.4	−15.7	−37.0
16 Bricks, pottery, glass, and cement	11.3	10.0	8.3	6.9	−11.3	−17.3	−16.8
17 Timber, furniture, etc.	7.3	5.8	5.5	5.4	−20.9	−5.0	−1.1
18 Paper, printing and publishing	9.5	10.0	11.8	11.5	+5.3	+18.1	−2.4
19 Other manufacturing	7.7	9.7	9.9	8.1	+26.1	+1.7	−18.3
Total manufacturing	**229.9**	**219.7**	**199.7**	**154.1**	**−4.4**	**−9.1**	**−22.9**
20 Total construction	**43.6**	**43.1**	**42.3**	**33.7**	**−1.2**	**−1.8**	**−20.4**
21 Gas, electricity, and water	10.9	12.7	12.1	11.5	+16.4	−4.2	−5.3
22 Transport and communication	41.1	35.8	33.7	31.2	−12.8	−6.0	−7.5
23 Distributive trades	82.5	76.7	80.7	75.2	−7.0	+5.1	−6.8
24 Insurance, banking, etc.	9.3	16.0	20.1	24.9	+62.6	+25.7	+24.3
25 Professional and scientific services	53.4	81.5	100.2	91.9	+52.6	+23.0	−8.3
26 Miscellaneous services	48.2	54.9	67.9	69.4	+13.9	+23.7	+2.2
27 Public administration[b]	26.3	44.9	54.7	49.2	+70.7	+21.8	−10.1
Total services	**272.3**	**322.5**	**369.4**	**353.3**	**+18.4**	**+14.6**	**−4.4**
Total all orders	**647.9**	**626.9**	**642.9**	**566.1**	**−3.3**	**+2.6**	**−11.9**

[a] 1961 data converted from the 1958 SIC by MLH disaggregation of 1958 order 4 (Chemicals etc.) into 1968 orders 4 and 5; and 1958 order 6 (Engineering and electrical goods) into 1968 orders 7, 8, and 9.

[b] 1961 figure does not include civil servants.

Source: MSC Census of Employment.

Table 4.2. *Male and Female Employment Change in the Newcastle Metropolitan Region by Industrial Sector, 1961–1981*

Industrial sectors (orders)		Employment (000s)				% Change in employment		
		1961[a]	1971	1978	1981	1961–71	1971–8	1978–81
Male employment								
1–2	Primary	98.8	39.7	29.9	23.7	−59.8	−24.8	−20.7
3–19	Manufacturing	166.6	159.8	148.1	115.3	−4.1	−7.3	−22.2
20	Construction	41.6	41.0	39.8	31.0	−1.5	−3.0	−22.1
21–27	Services[b]	138.2	151.1	161.1	153.8	+9.4	+6.6	−4.5
	Total	445.3	391.6	378.9	323.7	−12.0	−3.3	−14.6
Female employment								
1–2	Primary	3.3	1.9	1.5	1.3	−42.8	−20.9	−10.5
3–19	Manufacturing	63.3	59.9	51.6	38.8	−5.3	−13.9	−24.8
20	Construction	2.0	2.1	2.5	2.7	+4.0	+20.7	+5.9
21–27	Services[b]	134.1	171.3	208.4	199.5	+27.8	+21.6	−4.2
	Total	202.7	235.2	264.0	242.3	+16.1	+12.2	−8.2

[a] 1961 data converted from the 1958 SIC by MLH disaggregation of 1958 order 4 (Chemicals etc.) into 1968 orders 4 and 5; and 1958 order 6 (Engineering and electrical goods) into 1968 orders 7, 8, and 9.
[b] 1961 figure does not include civil servants.
Source: MSC Census of Employment.

ing and shipbuilding occurred within the broadly defined labour market area of the Newcastle metropolitan region, while other parts of the North East, notably Teesside, were relatively prosperous. Although not conceived in urban policy terms, many of the regional policy initiatives of this period significantly influenced the development of the conurbation (the designation of Washington New Town, for example). In addition, expansion of the service sector, especially public services, together with extensive city centre renewal, served to strengthen Newcastle itself as a regional capital.

Economic Change in the 1970s and 1980s

The period up to the mid-1970s represented, in many respects, a continuation of the perspectives and policies established at the time of the Hailsham Report: 'modernization', diversification through branch-plant operations, and service sector development. Total employment grew by 2.6 per cent over the period 1971–8 because further service sector growth was more than sufficient to compensate for continuing losses in manufacturing and mining. Manufacturing employment declined by 9.1 per cent over this period, largely owing to reductions in the engineering industries; the rate of decline was below that for Great Britain as a whole (−14.3 per cent) despite the structural characteristics of the area's economy. Regional policy helped to slow the rate of manufacturing job loss by supporting investment in some existing

Table 4.3. *Shift-Share Analysis of Employment Trends in Selected Metropolitan Regions, 1971–1978*

Region	Total employment change		Structural shift		Differential shift	
	(000s)	(%)	(000s)	(%)	(000s)	(%)
Newcastle	16.0	2.6	−3.8	−0.6	2.7	0.4
Glasgow	−30.2	−3.7	−4.0	−0.5	−48.5	−5.9
West Midlands	−10.6	−0.8	−24.4	−1.8	−22.6	−1.7
Bristol	17.4	4.2	2.0	0.5	4.2	1.0
London	−94.2	−1.6	197.3	3.4	−448.2	−7.8

Source: MSC Census of Employment.

industries and attracting branch plants: in 1978, 9.9 per cent of manufacturing jobs in Tyne and Wear were at relocated or branch-plant operations which had opened in the area since 1965 (Tyne and Wear County Council 1982). Job losses in mining over this period were smaller than over the previous decade since so much employment had already disappeared. As in the 1961–71 period, strong growth in public sector services (and also some private services, such as insurance and banking) largely generated female employment (much of which was part time) while losses in mining and manufacturing had a considerable impact on male employment. By 1978, only 31.1 per cent of jobs in the area were in manufacturing industry and only 4.5 per cent in mining; 41.1 per cent of employees were women. The area had undergone significant economic 'adjustment' and it was said to have 'successfully initiated a far-reaching process of restructuring and modernisation' (Northern Region Strategy Team 1977, 100).

Employment trends may be further examined through shift–share analysis: this provides a comparison between *actual* employment change and the change which would be '*expected*' if the local economy had followed national employment trends. Table 4.3 shows that, over the period 1971–8, the Newcastle metropolitan region had an overall employment growth of 0.4 per cent greater than would have been 'expected', given its industrial structure. This is accounted for by the contribution of public service growth which provided more than 9000 jobs over and above what would be expected had employment changed in line with national trends. The major differential loss, to offset these 'extra' jobs in public services, was in primary industries. Manufacturing performed little worse than 'expected' in relation to the national pattern of decline in those industries which made up the industrial structure of the study area. Overall, it appears that the economy performed relatively well in this period. This is further indicated by comparing the five metropolitan regions of the ESRC case-studies: only Bristol and Newcastle experienced an increase in total employment and they also both performed better than 'expected'. In spite of this, however, unemployment represented

Table 4.4. *Shift-Share Analysis of Employment Trends in Selected Metropolitan Regions, 1978–1981*

Region	Total employment change		Structural shift		Differential shift	
	(000s)	(%)	(000s)	(%)	(000s)	(%)
Newcastle	−76.1	−11.8	−1.8	−0.3	−41.6	−6.5
Glasgow	−77.7	−9.8	−5.1	−0.6	−32.3	−4.1
West Midlands	−134.1	−10.1	−60.8	−4.5	−5.6	−0.4
Bristol	−13.1	−3.1	12.7	3.0	−3.9	−0.9
London	−142.7	−2.5	170.1	3.0	−23.7	−0.4

Source: MSC Census of Employment.

an increasingly serious problem, both because service sector growth had not compensated for the *kinds* of jobs lost in the staple industries and because the labour force was increasing as a consequence of demographic trends and higher rates of female participation. Unemployment in the study area nearly doubled from 4.9 per cent in 1973 to 9.6 per cent at June 1979.

Economic conditions dramatically deteriorated from 1978–9 both nationally and locally. Over the period 1978–81, total employment in the area fell by 11.9 per cent, and this decline has since continued. As Tables 4.1 and 4.2 show, almost every industry experienced job losses, but the strongest impact was in manufacturing, which primarily affected male employment. In three years, manufacturing employment was reduced by 22.9 per cent, and by 1981 the manufacturing sector accounted for only 27.2 per cent of all jobs in the study area. The heavy traditional industries suffered the largest job losses, reflecting the long-term problems of lack of competitiveness and overseas competition, compounded by falling demand in conditions of recession. However, the newer lighter manufacturing industries have also been affected. Thus, between 1978 and 1981, the shift–share analysis revealed a 6.5 per cent decline of employment over and beyond that which would be 'expected' if the area's industries had followed the national trend—a considerably worse performance than those of the other case-study metropolitan regions (Table 4.4). Of the 41 600 'extra' jobs lost during this period, 14 600 were in manufacturing, 7400 in private services, and 10 600 in public services. Clearly, the policies pursued during the previous decade had not succeeded in creating a strong private sector economy that was capable of withstanding national recession, while public expenditure reduction undermined the growth in service employment which previously had compensated (at least numerically) for the decline of manufacturing jobs. One consequence of these trends was that unemployment in the study area reached 15.2 per cent by June 1981. Further job losses since 1981 have produced further increases in unemployment. At May 1985, the area's unemployment rate was 17.5 per cent compared with 13.2 per cent for Great Britain and 18.8 per cent for the Northern

Region. 92 180 males (20.8 per cent rate) and 35 471 females (12.6 per cent) were registered as unemployed *claimants* in the study area. These figures understate the level of unemployment since they exclude people ineligible for benefit, those on Manpower Services Commission (MSC) schemes, and those 'discouraged workers' who would enter the labour market if jobs were available.

To what extent are these high levels of unemployment associated with a mismatch of labour demand and supply between the inner areas, which had become the focus of urban policy, and other parts of the metropolitan region? Figure 4.2 demonstrates wide divergence between a very few areas which experienced employment growth between 1971 and 1981 and very many others affected by decline. The greatest percentage decline was in the Consett area following the steelworks closure. Several areas—notably Bedlington in Northumberland, East Boldon on the edge of Tyne and Wear, and also Sunderland—were affected by pit closures over this period, while inner areas within Tyneside and Wearside lost substantial proportions of jobs in engineering. Most of the older inner industrial and coalfield areas had rates of decline in excess of the average (−9.7 per cent) for the whole study area. Local factors, such as greater structural diversity and the growth of services, helped to reduce the rate of loss in some areas. In contrast, employment growth was confined to a few locations, notably the New Towns of Washington, Cramlington (Northumberland) and Killingworth, and the county towns of Morpeth and Durham. The New Towns, with green-field sites and industrial estates, succeeded in attracting a large share of branch-plant developments in the 1970s; they thus gained considerably from regional policy measures and also, as growing centres of population, acquired service sector employment. They thus contributed to the suburbanization of population and economic activity. Durham and Morpeth, both growing in population, gained services and some manufacturing industry. However, the pattern of employment change across the study area is not *simply* one of inner urban area decline and suburban growth, not least because of the influence of the coalfield and of Consett; nevertheless, this is a dimension of the distribution and one which, in the case of the New Towns, reflects the operation of public policy.

These differences in the pattern of job losses are, to a large extent, reflected by variations in unemployment rates within the metropolitan region. Figure 4.3 shows the spatial incidence of unemployment in May 1985 at ward level (amalgamated into zones). This indicates high rates in the older urban areas of Tyneside and Wearside designated as inner areas under the Inner Urban Areas Act.[1] This concentration of high unemployment stems from several interconnected factors which link deprivation and vulnerability in the labour market to the operation of housing markets: deprived residents in deprived areas are heavily dependent on job opportunities in declining industries within the areas of high job loss. All the designated inner areas recorded unemployment

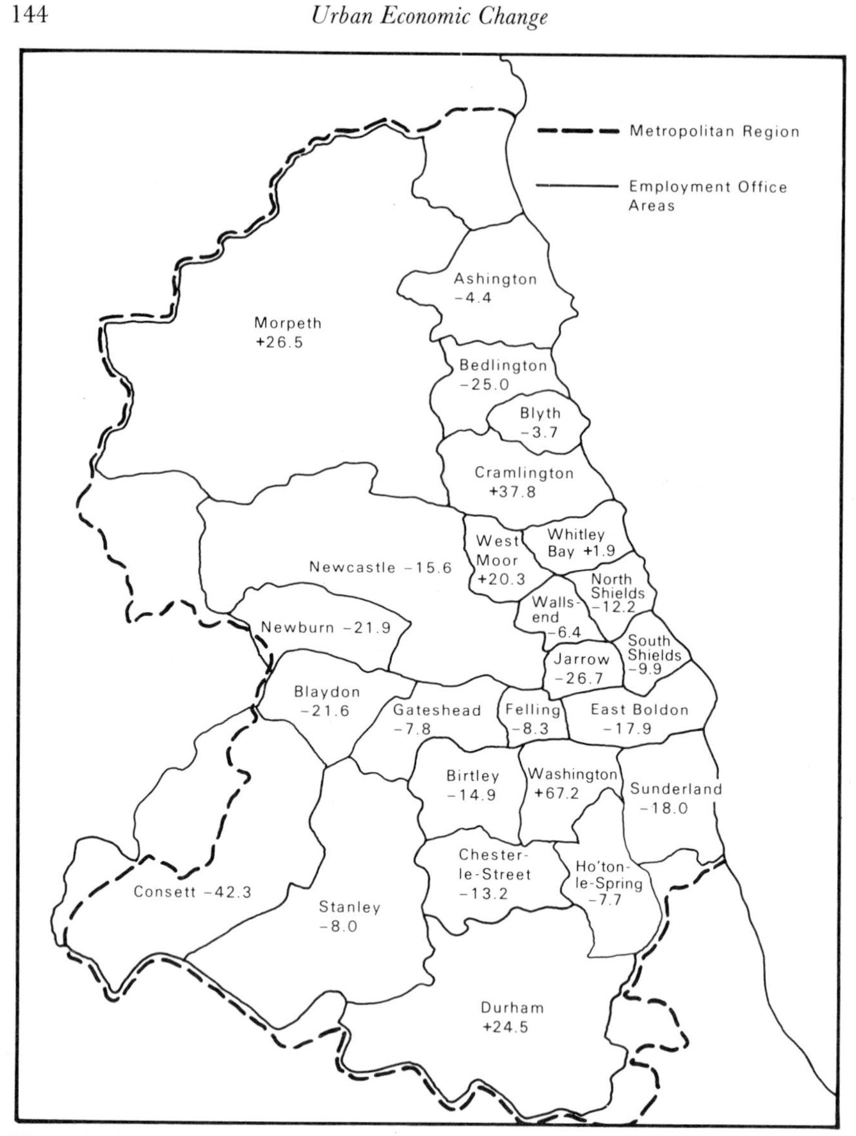

Fig. 4.2. Newcastle Metropolitan Region: Percentage Total Job Loss, 1971–1981, by Employment Office Area

Note: Study area boundary (defined by local authority districts), — — —.
Source: MSC Census of Employment.

rates above the average for the area as a whole (18 per cent) in May 1985, but five non-designated areas were also above the overall average rate. Analysis of unemployment data for smaller areas—postcode sectors—in Tyne and Wear (Tyne and Wear County Council 1985) shows that the localities which experienced the most acute levels of unemployment (at April

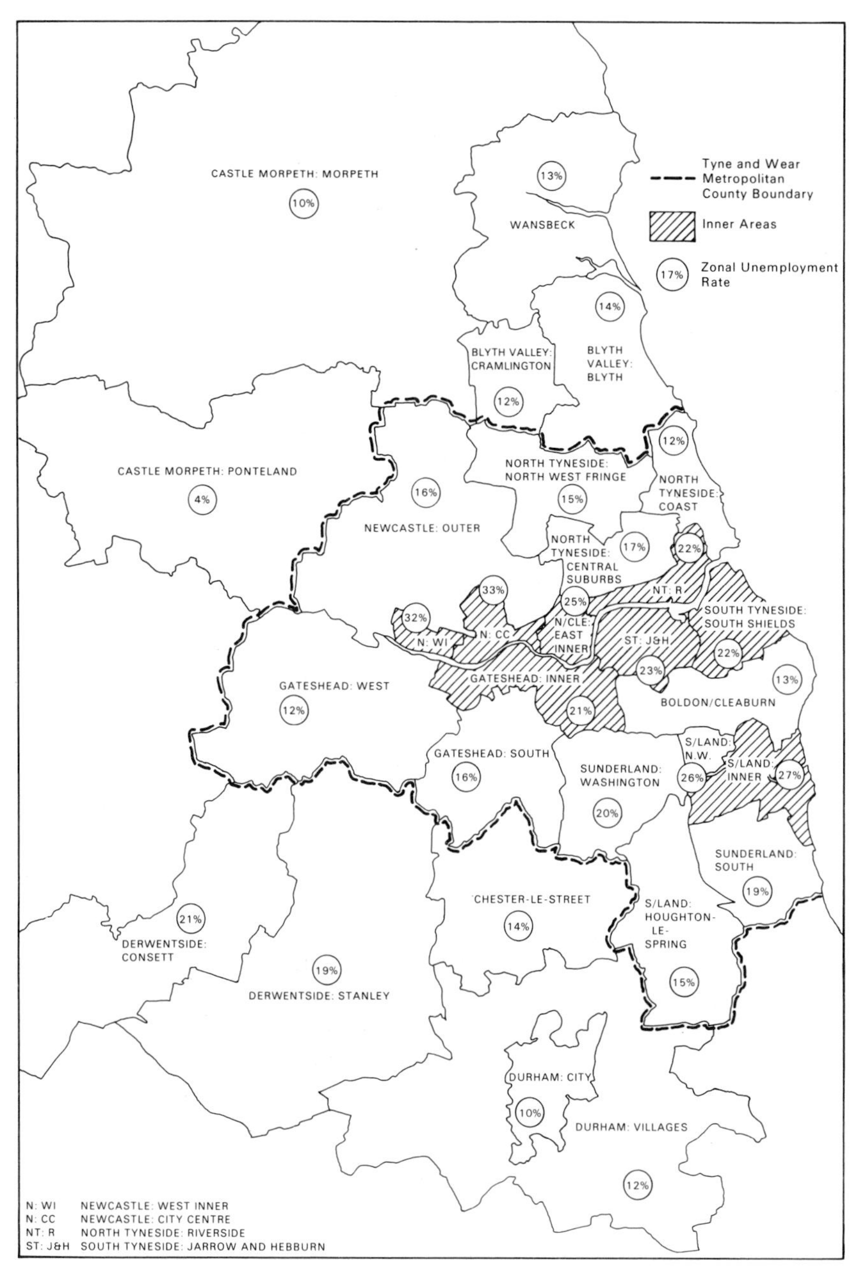

Fig. 4.3 Newcastle Metropolitan Region: Total Unemployment, May 1985
Source: Department of Employment.

1985) were mainly in the inner areas of the conurbation: the highest rate (47.3 per cent at April 1985) was recorded in Sunderland's East End, followed by Elswick Park in inner Newcastle, with a rate of 44.7 per cent. However, some peripheral estates also had very high rates—notably Newbiggin Hall estate in Newcastle (36.7 per cent at April 1985) and Thorney Close, Sunderland (31.8 per cent).

The high rates of unemployment (principally, but not exclusively, in the inner areas) reflect both the pattern of job loss in the area and the functioning of spatial labour markets. These high local rates of unemployment must be seen in the context of an overall shortfall in the demand for labour throughout the metropolitan region: it is clear that unemployment constitutes the most important economic, social, and political problem in the area.

The Structural Problems of the Metropolitan Economy

While we recognize that unemployment, economic decline, and de-industrialization are problems of the national economy as a whole, there are certain features of the Newcastle metropolitan region which may be detrimental or problematic in relation to its economic development and which have contributed to its above-average level of unemployment. Some of these features may be amenable to policy intervention, others not; but consideration of these issues serves to indicate possible objectives for policy and the limitations of policy.

Research undertaken since the late 1970s has established a degree of consensus about the major characteristics and weaknesses of the economy of the Northern Region, and these also generally apply to the Newcastle area or have particular implications for its development (Goddard 1983). Firstly, there remains a strong emphasis and dependence on heavy manufacturing and coal-mining—industries that are undergoing long-term decline and which have major difficulties competing with low-cost producers in world markets which are, in general, contracting. The manufacturing sector—now a relatively small part of the area's economy—continues to be dominated by engineering and shipbuilding: these industries still provided 47.6 per cent of manufacturing jobs in the study area in 1981, compared with 50.6 per cent in 1961. Regional policy has resulted in some diversification (and this includes diversification within engineering), but not sufficient to fundamentally change the economic structure or make it significantly less vulnerable to recession. One of the features of the economy, stemming from the limited extent of diversification, is that there is an under-representation of 'growth industries'. Information technology, biotechnology, and robotics—those industries which are expected to lead growth in the future—are not well represented in the area. There is not, for example, a nucleus of information technology production in the Newcastle area like 'Silicon Glen' in Scotland.

Secondly, about 80 per cent of manufacturing employment in the area is in establishments which have their headquarters elsewhere (Storey 1982). Much of the area's manufacturing industry is corporately marginal and involves routine production of mature products. Research and development (R&D) and other higher-level functions are, in general, under-represented and the 'branch-plant economy' is poorly integrated into the economy in terms of supply linkages: it is vulnerable to corporate restructuring operations and does not tend to provide a mechanism for the achievement of 'self-sustaining' growth. Regional policy which, during earlier periods of growth and economic restructuring, promoted the transfer of branches to the area has helped to generate the development of this 'branch-plant economy'.

Thirdly, the dependence on 'traditional' industries and externally controlled branch plants leads to an inferior performance in *net* job creation. Studies of manufacturing employment change in northern England (Storey 1985), and also in Canada and the United States of America, suggest that *gross* job losses can be spatially invariant, while differences in *net* job losses can be attributable to the ability of areas to create new jobs. The industrial and ownership structure of establishments in the study area appears to provide an unfavourable climate for the creation of new jobs, some of which might be generated by new businesses, but the vast majority of which would be expected to be created in existing established firms. A further symptom of the area's dependence on 'traditional' industries and externally controlled branch plants is the generally low rate of product innovation, especially in small and medium-sized firms. This implies that the longer-term competitiveness of some firms is threatened because of their relative failure to incorporate new technologies within products or develop new products and hence keep abreast of developments in their markets. Policy has only recently begun to address this issue directly, both through national-level industrial policy and local authority initiatives.

A fourth, but less significant, factor is that the area has a low rate of new firm formation. In 1982 and 1983, the North was in sixth place, in terms of formation rates, out of the eleven standard UK regions (Ganguly 1984). However, the failure rate of all new businesses is high—more than 30 per cent cease trading within three years of start-up—and their contribution to the creation of new employment is modest: in the North East, only about 4 per cent of manufacturing workers are employed in businesses started during the last ten years (Storey 1985). Formation rates are relatively low in this region, but this is perhaps of less importance than the differences in the characteristics of new businesses formed in this region in comparison with those businesses formed in some other parts of the UK. Because of the industrial structure of existing businesses, the relatively low levels of wealth, and the high proportion of blue-collar workers in this area compared with others, new businesses formed here are less likely to grow into large firms (Whittington 1984).

Fifthly, the region has not participated significantly in the national expansion of private producer services. The growth of external control, particularly through acquisition of local companies, has resulted in the removal of headquarters staff and a decline in demand from industry for local services, these being increasingly provided from corporate headquarters. In particular, financial services are poorly developed in the area, and this increases the difficulty for small companies in obtaining suitable packages of finance.

Finally, a high proportion of the area's employment is in the public sector—both the nationalized industries and public services. Decisions concerning investment and employment in this sector are taken by government in the context of efforts to reduce public expenditure. Moreover, with the exception of regional industrial policy, the distribution of resources through the expenditure of government departments is not explicitly planned so as to differentiate between regions: there is no explicit 'regional dimension' (NECCA 1984). This is a crucial issue since the impacts of public expenditure decisions are very substantial, yet these aspects of economic development do not form part of spatial economic policy.

Briefly stated, these are some of the key *local* features of the metropolitan economy; initiatives to strengthen this economy and deal with the general problems of unemployment and the specific problems of the inner areas need to be seen in this context. Without prejudging the issue, it is fair to say that policy intervention has been, and remains, limited in its application to the types of problems we have identified.

The rest of this chapter discusses the economic development policies operating in the area and then presents some salient findings from an impact assessment of direct financial assistance policies. We begin by examining policy development over the period 1974–84, establishing the basic mechanisms of policy and outlining the agencies involved. This includes figures for expenditure which indicate both the level of resources devoted to economic development policy in the study area and the relative significance, in resource terms, of each agency and policy measure. We focus specifically on financial assistance policies which constitute a major element of economic development expenditure. A unique set of data has been assembled on assistance cases, and this is used to examine the scale and incidence of expenditure. Following this, some of the findings from surveys of assisted establishments are presented, indicating the impact of assistance on the operation of firms and their employment.

Responses to Economic Change: Policy Proliferation, 1974–1984

Rising unemployment in the Newcastle metropolitan region promoted an extraordinary expansion of local economic development policy and a proliferation of agencies and initiatives, particularly after the onset of deepening recession in 1979. Regional policy continued to be the most significant element of public policy devoted to economic development in the area, but there was a shift in emphasis and interest at central government level in the mid-1970s away from regional problems towards urban and inner city problems. The current Urban Programme, expanded by the Department of the Environment in 1978, provided resources for economic development policy in Tyne and Wear (and other conurbations), making possible the much greater involvement by the local authorities in economic policy. The more adverse economic climate has resulted in a reduced amount of mobile industry and, hence, attention also shifted to indigenous development and policies aimed at helping small firms; this, too, stimulated the involvement of local agencies, especially the local authorities. Policy experiments at local level were mirrored by initiatives by central government, including the Enterprise Zones and Urban Development Grants. There are now numerous policy measures and agencies (particularly in the small business field), and a complex overlay of spatial policies has been created. Not surprisingly, there is confusion: some of the most effective agencies assisting small businesses are, consequently, those that are most adept at 'signposting' clients to other agencies and that know the 'system' (of agencies and measures) well enough to secure the optimal 'package' of assistance.

Regional industrial policy and selective financial assistance (DTI)

The basic structure of *regional policy* has not radically changed since the mid-1960s, although its spatial coverage has been altered and the financial and political commitment to policy has decreased. It continues to be the case that the prime objective of regional policy is to reduce disparities in employment between regions and to increase job opportunities in travel-to-work areas. A phased reduction in the geographical coverage of regional policy in 1980 and 1982 did leave all of the travel-to-work areas in the Newcastle metropolitan region with Assisted Area status. The government's review of regional policy in 1984 resulted in a further reduction in areal coverage in the North-East, such that the revised travel-to-work areas in the Newcastle metropolitan region and the Teesside conurbation are now the principal Assisted Areas in the Northern Region. However, the metropolitan area as an entity *has not* become the explicit focus for regional policy.

The legislative basis for regional policy is the 1972 Industry Act, and in England policy is administered by the Department of Trade and Industry (DTI). Nearly all applications for assistance from firms in the study area are handled and determined by DTI officials within the region: the DTI North East Regional Office at Newcastle deals with all but the largest selective financial assistance cases and also administers the Regional Development Grant Office at Billingham. The main thrust of policy is financial assistance to firms, and the most important instrument has been Regional Development Grant (RDG). The RDG scheme was revised in 1984, but during the study period it was payable automatically on nearly all new manufacturing investment (RDG is not available for small projects: individual items of plant and machinery costing less than £1000, or less than £500 in small establishments, and building costs below £5000 are not eligible). RDG payments have represented by far the largest part of financial assistance to industry in the Newcastle metropolitan region, with estimated expenditure (at cost prices) of nearly £229 million between 1974 and 1984 out of a total for Great Britain of more than £4000 million. The next-largest source of DTI aid is Regional Selective Assistance (RSA), which is normally available in addition to RDG and which is used to support projects which create or safeguard jobs; applicants must demonstrate that it is necessary to enable the project to be implemented. Broadly, all projects in manufacturing, construction, and mining may be eligible, but assistance may also be offered for certain service sector projects. Our analysis reveals that over £86 million was granted in RSA to firms in the study area over the period 1974–84.

The DTI also operates *national schemes* under Section 8 of the Industry Act. These include National Selective Assistance (NSA) to support firms in the 'national interest' (it was used to help BL and Chrysler in the 1970s); general schemes to bring forward investment; technology schemes ('Support for Innovation' measures); sectoral schemes for specific industries and small-firms loan guarantees. Between 1974 and 1984, assistance from these schemes to firms in the study area totalled £11.5 million. The DTI has also provided financial aid to shipbuilding: figures for the study area are not available, but committed expenditure on the largest element, the Intervention Fund, amounted to £251 million by 1983 and perhaps a quarter of this went to yards on the Tyne and the Wear.

The incidence and impact of these (and other) financial assistance measures is discussed in the next main section of this chapter; for the moment, it is worth noting that most DTI assistance over this period went to support investment in manufacturing industry, that the majority of it is an automatic subsidy not linked to job creation or retention (but new criteria for RDG introduced such a link in 1984), and that there is no difference in treatment between 'inner city' areas and other parts of an Assisted Area.

In addition to the provision of financial assistance, the DTI plays an important part in economic development through a variety of policies operating nationally: these include services aimed at small firms; technical services; assistance to exporters; and support for tourism projects. The DTI also supports English Estates, which is responsible for factory construction in the English Assisted Areas. The North of England Development Council, which carries out industrial promotion for the region as a whole, is also supported by a grant from the DTI, together with subscriptions from the local authorities.

Local authority economic development policies

All the *local authorities* in the area are now involved in economic development policy. In some cases this is minimal and includes little more than small-scale provision of factory units and industrial sites (Castle Morpeth District Council); in others it comprises a comprehensive Economic Development Policy and Programme (Newcastle City Council). The most active authorities—which have the larger resources and a clear legislative basis for policy—are the metropolitan district councils and the county council in Tyne and Wear.

Comparable data on economic development expenditure by the local authorities are not available on a comprehensive basis since authorities differ in their interpretations of what constitutes economic development. However, the North of England County Councils Association (NECCA) has been assembling reasonably consistent information since 1979 on authorities' expenditure on the major components of economic development (Figures 4.4 and 4.5), and this serves, at least, to indicate broad patterns and trends. Figure 4.4, showing total expenditure over the period 1979–84, points to the contrast between the Tyne and Wear authortities and those in Northumberland and Durham. It also shows that, in relation to population, even the expenditures of the Tyne and Wear District Councils are fairly small at £22–39 per head over five years. Between 1979 and 1984 local authorities in the area spent a total of £56 million on a whole range of economic development activities; because of the way in which the NECCA data were compiled this is an underestimate, but, to place this figure in context, it does represent a level of expenditure amounting to less than half DTI spending on RDG (£129 million) over this period. Figure 4.5 indicates the categories of expenditure for all the local authorities, showing the overall increase since 1982 as policy became more established and increasing attention was paid to economic development, and as the emphasis in the Urban Programme shifted more towards economic projects. The table does show the considerable expansion of expenditure on grants to industry; but it also demonstrates the continued emphasis placed on the 'traditional' areas of local economic development policy, particularly the acquisition of land and construction of factories.

Authorities in *Tyne and Wear* first obtained significant powers to engage in economic development through the 1976 Tyne and Wear Act, successfully promoted as a local Act of Parliament by the county council and providing powers to both the county and district authorities. Subsequently, the 1978 Inner Urban Areas Act extended similar powers to designated local authorities in many parts of the country, and enabled additional resources to be directed to these areas from the revised and expanded Department of the Environment Urban Programme. The metropolitan districts of Newcastle and Gateshead were designated 'Partnership Authorities' (although the area over which the Partnership operates covers only a part of these two districts), and the other three metropolitan districts as 'Programme Authorities'. Within these areas, Urban Programme resources are concentrated within the defined 'inner areas'.

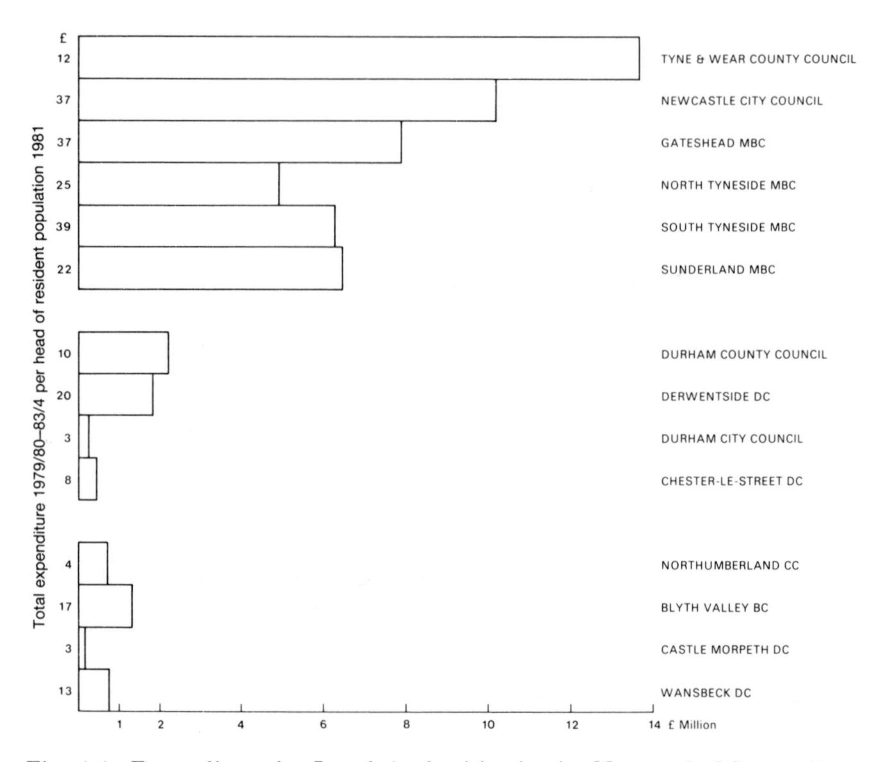

Fig. 4.4. Expenditure by Local Authorities in the Newcastle Metropolitan Region on Economic Development, 1979–1980 to 1983–1984

Notes: 1. Expenditure by Durham and Northumberland county councils apportioned to study area according to population.
2. Returns were not made for the following: Derwentside District Council (1981–2), Chester-le-Street (1979–80, 1983–4), Northumberland County Council (1979–80), Blyth Valley Borough Council (1979–80, 1980–1, 1983–4), Castle Morpeth District Council (1979–80), and Wansbeck District Council (1979–80).
3. The figure for Blyth Valley Borough Council includes one loan of £900 000.
4. Figures include expenditure made for Industrial Improvement Area purposes.
Source: NECCA.

The government's objectives for urban policy as a whole, stated in the 'Ministerial Guidelines' (Department of the Environment 1985), are:

(*a*) To improve employment prospects in the inner cities, by increasing both job opportunities and the ability of those who live there to compete for them
(*b*) To reduce the number of derelict sites and vacant buildings

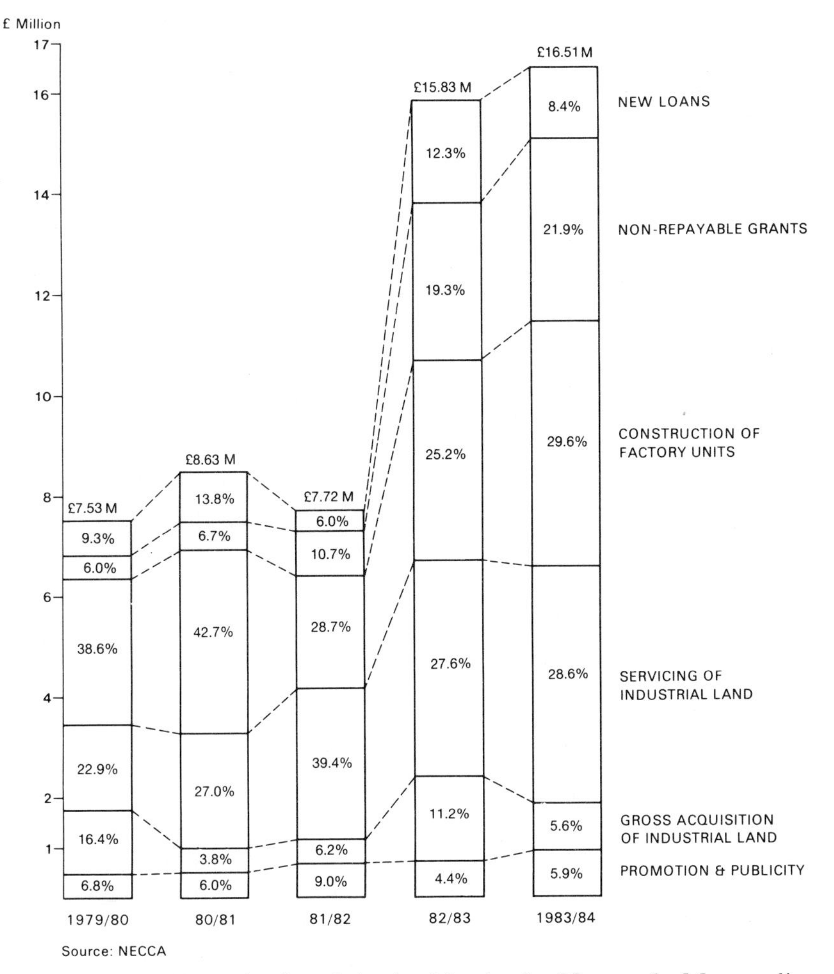

Fig. 4.5. Expenditure by Local Authorities in the Newcastle Metropolitan Region on Economic Development, by Category, 1979–1980 to 1983–1984

Note: Coverage of data as in Figure 4.4; see notes 1.2, 3, and 4 to that figure.
Source: NECCA.

(*c*) To strengthen the social fabric of the inner city and encourage self-help
(*d*) To reduce the number of people in acute housing stress.

For both the Partnership and the Programme Authorities an Inner Area Programme (IAP) is agreed with the Department of the Environment each year, comprising 'economic', 'environmental', and 'social' projects which will receive 75 per cent funding from the Department of the Environment, the local authority contribution being largely provided from Main Programme budgets or through the 2p rate mechanism of Section 137 of the Local

Government Act 1972. The Newcastle–Gateshead Partnership had a total IAP expenditure of £22.2 million in 1983–4 and the three Programme Authorities had a combined expenditure of £12.5 million. Within the Urban Programme an increasing proportion of expenditure has been directed to 'economic' projects, and by 1983–4 more than 45 per cent of all IAP expenditure in Tyne and Wear was in this category (Figure 4.6). Total Urban Programme expenditure on economic projects in Tyne and Wear over the period 1979–84 is shown in Figure 4.7, indicating the higher level of resources directed at the Partnership Authorities. These figures are not directly comparable with those compiled by NECCA (Figures 4.4 and 4.5), again because of definitional problems. The main point to be stressed is that the Urban Programme makes a substantial contribution to the economic development activities (however defined) of the Tyne and Wear local authorities. For example, in 1984–5, 46.9 per cent of budgeted expenditure in Newcastle City Council's Economic Development Policy and Programme formed a component part of the Partnership IAP and 33.6 per cent of Tyne and Wear County Council's Economic Development Committee expenditure was within the Partnership or Programme IAPs. Of this IAP expenditure, 75 per cent is funded by the Department of the Environment.

The acquisition and preparation of industrial sites and the provision of factory units have dominated the economic development programmes of the Tyne and Wear local authorities. Site provision includes acquisition for local authority factory construction and also preparation of strategic sites to attract mobile industry. Local authority factory provision has mainly involved the construction (or refurbishment) of small low-cost rented units for small businesses to meet perceived unfilled demand. In recognition of the needs of new businesses, some schemes have incorporated business support services (for example, Newcastle City Council's Enterprise Workshops). Expenditure on loans and grants to industry has also been aimed largely (and increasingly) at small and new businesses, and several authorities have simple grant schemes for new starters, while some provide additional support for people starting a business under the MSC Enterprise Allowance Scheme. In addition, four of the Tyne and Wear authorities now operate wage subsidy schemes to encourage recruitment of the unemployed, and these are financially aided by the European Social Fund. Since a large proportion of the resources for local economic development policy emanate from the Urban Programme, much of the expenditure is targeted on the designated inner areas in accordance with the practice of determining agreed IAPs. Much of the effort and expenditure is concentrated in the twenty-three Industrial and eleven Commercial Improvement Areas in the older inner areas of the county, where the aim is to improve operating conditions, restore business confidence, and secure employment through environmental and infrastructural schemes, together with the provision of factory units.

It is important to note that the majority of local authority spending con-

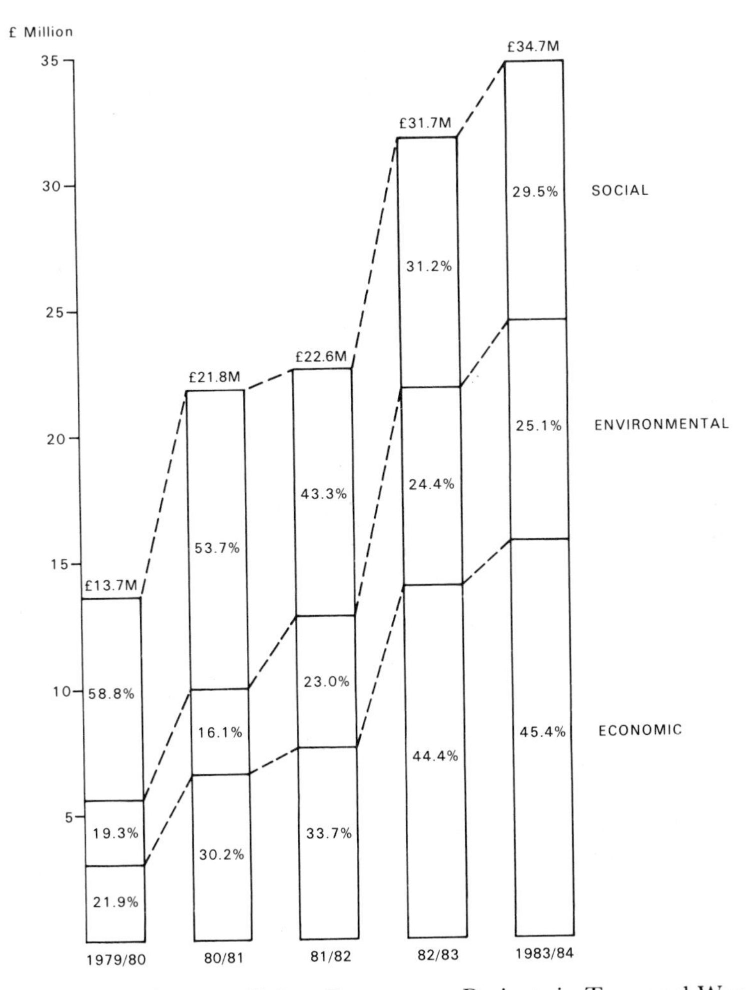

Fig. 4.6. Expenditure on Urban Programme Projects in Tyne and Wear, by Category, 1979–1980 to 1983–1984

Note: Actual expenditure on projects in the IAPs; 75 per cent of the expenditure is met by the Department of the Environment, the rest by the local authorities. This covers spending by both county and district authorities. The percentage of total expenditure devoted to each category in each year is also shown.

Source: IAPs and relevant local authorities.

tinues to be on sites, factories, and loans and grants to firms; the authorities have not, for example, followed the Greater London Council and West Midlands County Council in setting up an enterprise board to provide equity finance. Less conventional or more experimental approaches to economic development policy have tended to involve the funding of other agencies and

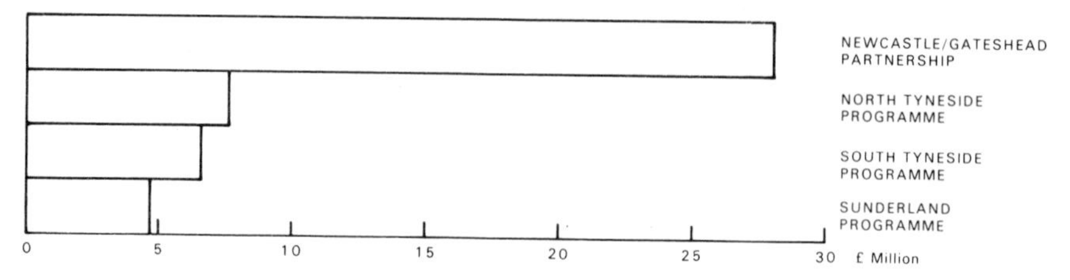

Fig. 4.7. Expenditure on Economic Projects in Tyne and Wear under the Urban Programme, 1979–1980 to 1983–1984

Note: The figures shown refer to actual expenditures on economic projects forming part of the IAPs approved by the Department of the Environment. 75 per cent of the expenditure is met by the Department of the Environment, the remainder by the local authorities. The IAPs cover economic development spending by both county and district authorities.

Source: IAPs and relevant local authorities.

the development of working arrangements with them. Thus, Tyne and Wear County Council aims to foster technological development and transfer by financially supporting the Micro-Electronics Applications Research Institute and the Tyne and Wear Innovation Centre; similarly, the county and other authorities help to finance the Northern Region Co-operatives Development Association.

The economic development activities of the local authorities in *Northumberland and Durham* are largely reliant upon authorities' own resources, mainly obtaining powers through the 2p rate mechanism of Section 137. Expenditure by these district and county authorities within the study area is estimated at £7.7 million over the five years 1979–84, or about the same amount as was spent by Gateshead alone over this period. Provision of small rented factories, sites, and some loans and grants to small firms are the main components of expenditure and there are few resources for experimentation. The contrast between the wide range of policies operated by Tyne and Wear authorities and the more limited range of policies in Northumberland and Durham is evident from Table 4.5.

Almost all local authorities in the study area engage in some promotion and publicity, ranging from participation in trade fairs and marketing campaigns to the publication of local business registers. The promotion of the local authority area as a location for inward investment was, for most authorities, the first economic development activity they established; now it is of much less importance.

Urban Development Grant and Enterprise Zone assistance (Department of the Environment)

In addition to the funding of assistance schemes submitted through the IAP exercise, the Department of the Environment provides resources enabling

Table 4.5. *Economic Development Policies of Local Authorities in the Newcastle Metropolitan Region, as at 1985*

Policy	Tyne and Wear						Co. Durham				Northumberland			
	Tyne and Wear CC	Newcastle	Gateshead	N. Tyneside	S. Tyneside	Sunderland	Durham CC	Chester-le-Street	Derwentside	Durham City	Northumberland CC	Castle Morpeth	Blyth Valley	Wansbeck
Acquisition and servicing of industrial land	•	•	•	•	•	•	•		•		•	•	•	•
Industrial/commercial improvement areas	•	•	•	•	•	•								
Construction of factory units	•	•	•	•	•	•	•	•	•	•	•	•	•	•
Managed workshops	•	•	•	•	•	•								
Loans and grants	•	•	•	•	•	•	•	•	•	•	•		•	•
Wage subsidies	•	•			•	•	•		•				•	•
Promotion and publicity (inward investment)	•	•	•				•							
Tourism promotion	•	•			•		•			•	•			
Advisory services	•	•				•	•	•	•		•		•	•
Technology initiatives	•	•		•		•								
Support for co-operatives	•	•		•	•	•				•	•			
Support for community enterprise	•	•				•								

Sources: Committee minutes and discussions with local authority officers.

financial assistance to be given to firms through two further measures: *Urban Development Grant* (UDG) and *Enterprise Zone* concessions. UDG is a grant scheme administered by local authorities which is designed to 'promote the economic and physical regeneration of inner urban areas by levering private sector investment into such areas' (Department of the Environment 1983). From its inception in 1983 up to June 1985 £5.8 million of UDG, levering nearly £50 million of private sector investment, was offered to support twelve schemes in the designated inner areas of the Newcastle metropolitan region, involving commercial development (two schemes) and housing refurbishment (ten schemes). A further eight schemes have been offered UDG but have not yet started. Bids for UDG support are submitted by the local authorities, in conjunction with private developers, to the Department of the Environment. Proposals are then assessed by both the regional and national offices of the Department. If a proposal is successful, the Department of the Environment contributes 75 per cent of the UDG support, the remainder being funded by the local authority submitting the application.

The Newcastle–Gateshead Enterprise Zone, set up in August 1981, comprises old industrial land along the banks of the Tyne together with the southern section of English Estates' Team Valley Industrial Estate, and firms within the zone receive a subsidy in the form of exemption from local authority rates and other incentives (this provision is to operate for ten years from the date of designation). In the period since designation up until March 1984, the exemption from rates in the Newcastle–Gateshead Enterprise Zone amounted to nearly £7.3 million. These amounts are reimbursed to local authorities by central government.

Assistance from the European Economic Community

Finally, resources for economic developement in the study area are also provided by the European Economic Community. Between 1975 and 1984, an estimated £60 million was allocated to the area from the *European Regional Development Fund*, (ERDF) four-fifths of which helped to finance infrastructure projects (such as the Metro rapid transit system) while the rest was allocated to industrial projects; however, ERDF aid is not additional, but serves as a refund to government for DTI regional assistance spending. Tyne and Wear is also eligible, as an area affected by the decline of shipbuilding, for funding from the ERDF 'non-quota' section. This provides help for small and medium-sized enterprises through the provision of premises, infrastructure, and support for advisory agencies. The bulk of *European Social Fund* (ESF) expenditure in the area is provided as reimbursement to government for spending on MSC activities, but the ESF also supports some specific local initiatives, notably the Tyne and Wear Enterprise Trust (Entrust) and wage subsidy schemes. Entrust offers business training and counselling, promotes self-employment, and also serves some local authorities in vetting grant applications; in addition to the ESF, it receives some funding from the Urban Programme. The ESF provides 30–50 per cent of the finance for wage subsidy schemes which provide employers with a subsidy if recruiting the young or the longer-term unemployed. ESF expenditure on these schemes since they were introduced in 1981 up to March 1984 amounted to almost £1 million. A third source of funding comes from the *European Coal and Steel Community* (ECSC) which offers 'readaptation grants' to help fund early retirement, retraining, and income support measures in areas suffering job losses in coal and steel, and also provides low-interest loans to companies expanding in these areas. An estimated £40 million from the ECSC came to the study area between 1977 and 1984.

Other agencies and policies

Public sector provision of factory premises has been a major part of regional policy since the 1930s. *English Estates* (which operates as an agent of the DTI and has its regional and national headquarters at Gateshead) and *Washington Development Corporation* (responsible to the Department of the Environment) have traditionally built large advance factories, mainly at green-field sites on the conurbation periphery (though English Estates has increasingly built within the inner areas). In recent years, both have built smaller units, complementing the more recently established programmes of the local authorities. English Estates and Washington Development Corporation offer financial assistance to tenants of their factories in the form of rent relief: between 1974 and 1984 rent relief to firms in the study area from English Estates amounted to £5.7 million and Washington's tenants received assistance of more than £2 million (partially refunded by the DTI). Washington Development Corporation is also active in industrial promotion and publicity, both in the UK and overseas, to support the economic development of the New Town.

Most financial assistance to firms is provided in the form of grants or, less often, secured loans. However, public sector aid through the provision of equity finance was formerly available from the *British Technology Group* (BTG). BTG incorporated the National Enterprise Board set up in 1976, and, until its equity function was abolished in 1984, its objective was to provide and arrange development capital for viable industrial projects which had difficulty in obtaining private sector support. Between 1976 and 1984, BTG (which had a Newcastle office serving the North East as a whole) invested £5.6 million in firms in the study area which, according to BTG, 'levered' six times that amount from private sources. These investments are now gradually being relinquished.

Throughout the area there are a number of agencies (including several local authorities) involved in helping to arrange assistance packages and provide counselling services to small firms and intending entrepreneurs. Among these is Derwentside Industrial Development Agency, an enterprise agency which has now taken over the role of *British Steel Corporation (Industry) Ltd.*—BSC(I)—in the Consett–Derwentside area. The agency provides advice, draws up business plans, and helps to negotiate assistance from other sources. Some aid is directly available from BSC(I) in the form of loans and grants: such assistance was paid to 130 applicants in Derwentside between 1979 and 1983, amounting to a total of £1.4 million.

Within the Newcastle metropolitan region there are many other agencies involved in economic development, most of which receive some public sector support, often from the Urban Programme and the local authorities. Here we have sought merely to outline some of the *main* agencies, programmes, and sources of financial aid; this account is not comprehensive, but indicative. Moreover, a broader definition of economic development would bring in other agencies and policies, perhaps the most important of which is the MSC. In 1983–4, MSC expenditure in the study area was estimated at £72 million—considerably more than all the other sources of expenditure combined; most of this was spent on the Youth Training Scheme and the Community Programme, but resources (of £3.4 million in 1983–4) were also spent on the Enterprise Allowance Scheme which supports new business starters.

Financial assistance expenditure: an overview

Because of the great diversity of measures involved—from infrastructure improvements, training, advisory services, loans, and grants, many of which have only an indirect influence on the economic performance of firms and/or job creation—it is not possible to bring all of the preceding figures together and derive a meaningful comparison of economic development expenditure by the different agencies. However, data on *financial assistance expenditure*, on a case-by-case basis, have been collected from the main assistance agencies; and, from this information, it has been possible to examine the scale and characteristics of assistance expenditure. Table 4.6 indicates the scale of assistance from the various sources, revealing that a total of nearly £370 million was paid out in assistance to firms in the Newcastle metropolitan region over the period 1974–84. In terms of the agencies involved, the most important point to emerge from this table is the dominance of DTI expenditure which accounted for 88.2 per cent of the total. RDG was the major policy instrument, representing 61.8 per cent of all assistance expenditure.

At first sight, the sheer scale of resources involved appears impressive but, in fact, it is not large in relation to the size of the problem: the Exchequer cost of unemployment in the area is estimated to be in the region of £750 million a year (Sinfield and Fraser 1985), while the total economic development expenditure of the DTI and the local authorities (including the Urban Programme) is now some £35 million a year. Moreover, the resources made available through the Urban Programme should be considered in the context

Table 4.6. *Direct Financial Assistance to Industry in the Newcastle Metropolitan Region, 1974–1984, by Agency*

Agency	Expenditure	
	(£m.)	(%)
DTI—Of which:	326.2	88.2
RDG[a]	(228.4)	(61.8)
RSA	(86.3)	(23.3)
NSA[b]	(11.5)	(3.1)
Local authorities[c]	14.9	4.0
English Estates	5.6	1.5
BTG	6.4	1.7
Washington Development Corporation[d]	2.4	0.7
BSC(I)[e]	1.4	0.4
UDG	5.8	1.6
Enterprise Zone	7.2	1.9
All agencies	369.9	100.0

[a] Includes apportionment for RDG payments below £25 000 which are not known.
[b] Excludes NSA assistance given under shipbuilding schemes and also the self-financing Small Firms Loan Guarantee Scheme because DTI data were not made available.
[c] Includes payments made to firms for Industrial and Commercial Improvement Area purposes. Coverage not totally complete.
[d] Excludes some pre-1978 cases.
[e] Excludes assistance in 1983–4.

Source: British Business and relevant agencies.

of reduced rate-support grant paid to the local authorities: the authorities maintain that they have 'lost' substantially more through reductions in rate-support grant in recent years than they have gained from the Urban Programme.

We previously noted that there are considerable difficulties in attempting to evaluate the impact of economic development expenditure: indeed, it has been a major task of the research merely to obtain figures on amounts spent and the projects or recipients involved. The next section presents details on patterns of expenditure on direct financial assistance and discusses the impact and effectiveness of this assistance in relation to the problems of the area.

The Incidence and Impact of Financial Assistance

Three main approaches have been adopted in this study to establish the impacts of direct financial assistance to industry:

(*a*) Agencies providing direct financial assistance to firms in the area were asked to supply detailed data on individual cases of assistance over the period 1974–84. This covers, on a case-by-case basis: amounts of assistance; type and purpose; date; industrial classifications and location of the establishment. Information on 2457 establishments known to have received assistance was collected and a comprehensive computerized data base has been assembled.

(*b*) A detailed sample survey was conducted of managers of assisted establishments. The survey covered a stratified random sample of establishments drawn from the data supplied by the assistance agencies themselves. The sample comprised establishments assisted over the last four years—this was considered a suitable time-span to obtain reliable data on assisted projects and their outcomes. Altogether, 201 interviews with managers were successfully completed.

(*c*) To explore labour market impacts, a questionnaire survey of employees was carried out at establishments which had been included in the survey of managers. Questionnaires were completed and returned by 1388 employees at 120 of the assisted establishments, producing a response rate of 44 per cent. The questionnaire elicited information about the location of employees recruited by assisted establishments, their occupations, skills, and other characteristics to determine how the benefits of employment created in assisted firms are distributed within the labour market.

Here we draw on some of the information from these sources to assess the extent to which assistance measures have addressed the structural problems of the area and served to generate employment. Four main themes are explored below: industrial diversification; size structure; employment effects; and cost-effectiveness.

Industrial diversification

The dependence of the metropolitan region on nationally declining industrial sectors and the relative underdevelopment of services have been noted as basic structural weaknesses of the local economy. A related weakness has been a low rate of product innovation within manufacturing firms which might contribute to industrial diversification. To what extent has financial assistance been used to support industries experiencing long-term employment decline and how far has it been directed towards those activities which are experiencing growth or have good growth prospects? Furthermore, to what extent has the provision of assistance supported diversification within existing firms through the introduction of new products?

An analysis of assistance by industrial sector reveals that the overwhelming majority of the aid has gone to manufacturing. Of all expenditure that could be classified in terms of the economic activity of the recipient establishment, 88.0 per cent was paid to manufacturing firms. This is basically a reflection of the orientation of DTI policy which accounts for the bulk of the assistance. However, it is the case that all other agencies for which infor-

Table 4.7. *Direct Financial Assistance in the Newcastle Metropolitan Region, 1974–1984: Sectoral Analysis of DTI and Local Authority Assistance within Manufacturing Industry*

Industrial sectors	% Manufacturing employment 1978	% DTI assistance	% local authority assistance	Quotient[a]	
				DTI	Local authority
Food, drink, and tobacco	7.2	11.0	4.6	1.53	0.63
Coal, petroleum, and chemical products	6.0	15.8	7.3	2.63	1.21
Metal manufacture	6.0	8.0	5.4	1.30	0.90
Mechanical, electrical, and marine engineering	53.3	49.5	39.4	0.92	0.74
Textiles and clothing	9.7	3.8	8.9	0.39	0.92
Other manufacturing	17.8	11.8	34.2	0.66	1.92

[a] Quotient calculated as (% assistance) ÷ (% total manufacturing employment)—that is, (column 2 or 3) ÷ (column 1).

Sources: MSC Census of Employment, *British Business*, and relevant agencies.

mation is available also concentrate their expenditure on manufacturing. Even local authorities which proportionately provide the most assistance to the service sector (22.9 per cent) directed 72.9 per cent of their assistance towards the manufacturing sector. This can be compared with manufacturing employment, which in 1981 accounted for 27.2 per cent of the study area's total employment.

Table 4.7 shows that within the manufacturing sector the pattern of assistance does not mirror the existing industrial structure. Food, drink, and tobacco; coal, petroleum and chemical products; and metal manufacture—all are disproportionately represented in the DTI assistance, while 'other manufacturing' features prominently in the case of local authority assistance. This difference partly reflects the orientation of RDG payments towards capital-intensive projects. What is more striking is the concentration of assistance in a few Minimum List Headings (MLHs) of the Standard Industrial Classification (SIC) (Table 4.8). In the case of DTI assistance, the ten leading MLHs (out of a total of ninety-nine MLHs in manufacturing with more than a hundred employees in the area in 1978), received 54.8 per cent of the assistance. These ten MLHs accounted for 34.2 per cent of total manufacturing employment. In the case of local authority assistance, the top ten recipient MLHs in manufacturing accounted for 46.5 per cent of aid and only 26.0 per cent of manufacturing employment. Hence, for both DTI and local authority assistance, there was a high level of concentration of expenditure on a few MLHs. Local authority assistance within manufacturing was therefore not significantly less concentrated in terms of specific industries than the DTI assistance.

Table 4.8. *Top Ten Manufacturing Industries Assisted by the DTI and Local Authorities in the Newcastle Metropolitan Region, 1974–1984*

Industry (MLH)	% of Assistance	% Manufacturing Employment 1978	% Employement change 1971–81	
			NMR[a]	GB
DTI				
Shipbuilding	17.8	12.6	−21.8	−21.5
Pharmaceuticals	6.1	0.9	−15.9	−0.3
Other machinery	5.9	4.2	−9.7	−25.3
Brewing	5.3	1.8	−14.5	−13.2
Other mechanical engineering	4.2	3.5	−27.3	−10.5
Radio and electronic components	4.0	3.5	+15.7	−15.5
Iron and steel	3.9	3.6	−78.7	−50.1
General chemicals	3.2	1.7	−28.2	−13.4
Construction equipment	2.3	1.5	−10.9	−30.7
Aluminium manufacture	2.1	0.9	+149.7	−20.1
Total	54.8	34.2	—	—
Local authorities				
Other printing and publishing	10.0	3.2	+43.7	+3.5
Other mechanical engineering	8.2	3.5	−27.3	−10.5
Synthetic resins and plastics	4.9	0.9	−60.7	+3.0
Motor vehicle manufacture	3.7	1.1	−36.9	−31.5
Shipbuilding	3.7	12.6	−21.8	−21.5
Metal industries n.e.s.	3.2	2.4	−26.4	−28.7
Timber	3.2	1.0	+0.6	−30.2
Wire manufacture	3.2	0.6	−55.0	−37.2
Miscellaneous manufacturing	3.2	0.5	−76.5	−26.8
Paper and board	3.2	0.2	−84.5	−39.7
Total	46.5	26.0	—	—

[a] NMR, Newcastle metropolitan region.

Sources: British Business, the DTI, local authorities, and the Department of Employment.

Table 4.8 shows a more marked contrast in terms of the particular industries (MLHs) receiving assistance from the DTI and local authorities; the Table also shows employment change in these industries as a whole (that is, all establishments, whether assisted or non-assisted) in the study area and also in Great Britain. The traditional declining sectors of the area, notably shipbuilding and steel, received a significant amount of assistance from the DTI, but the DTI also gave a large amount of support to newer sectors, such as pharmaceuticals and radio and electronic components. The industries which received a large share of local authority assistance also included several metal-based activities, but a significant amount of assistance was also given to 'low-technology' areas, like timber.

A further indicator of the sectoral concentration of financial assistance can be given by identifying the ten individual establishments most heavily assisted over the whole ten-year period. Unfortunately, because most local authority data were given to us on a confidential basis we cannot list the leading local-authority-assisted establishments; the principal recipients of DTI aid are, however, published, and these are given in Table 4.9. The Table indicates how the principal heavy engineering companies of the area—British Shipbuilders and Northern Engineering Industries PLC (NEI)—have benefited from regional and sectoral aid schemes. Brentford Nylons (which was 'rescued' with the support of RSA), the breweries, Sterling Winthrop, Boots, and Welwyn Electric were the only significant recipients of support outside the engineering sector.

While financial assistance may not be diversifying the industrial structure at the aggregate level, it does appear to have supported projects leading to change in products *within* existing industrial establishments. Responses to our survey of managers in assisted establishments showed that 55.2 per cent of those projects receiving RDG involved the introduction of new products. In the case of RSA/NSA the proportion was higher, at 75.8 per cent, but lower for projects supported by local authority assistance (48.0 per cent). In only 26.7 per cent of RDG and 28.0 per cent of local authority cases did assisted projects involve an increased R&D effort, although the figure was significantly higher in relation to projects supported by RSA/NSA (45.5 per cent). In the majority of instances, financially assisted projects were reported to have led to increases in productivity, profitability, and output—beneficial impacts in terms of improving the operating performance of firms in the area.

The size distribution of assisted establishments

Another feature of the local economy, which is related to the industrial structure, is the low proportion of manufacturing employment in small firms, coupled with a below-average rate of new firm formation. Many of the economic development policy measures of the local authorities and other agencies have been aimed at fostering a high rate of new firm formation and providing advisory services for new small businesses. To what extent has *financial assistance* been directed towards the small-firm sector?

Table 4.10 shows the size groups, and average and median employment, of assisted establishments. Over half of the *cases* of assistance went to establishments employing 1–25 people. Even allowing for deficiencies in the data (notably the exclusion of small RDG payments), the contrast between agencies—especially between the DTI and the others—is substantial. The DTI has tended to assist mainly the larger establishments; the median employment size groups of the recipients of (larger, published) RDG payments was 201–500, while for RSA and NSA it was 51–100. Other agencies, with much smaller resources, concentrated their efforts on much smaller

Table 4.9. *The Ten Establishments in the Newcastle Metropolitan Region Most Heavily Assisted by the DTI, 1974–1984*[a]

Rank	RDG[b]	RSA[b]	NSA[b]	RDG, RSA, and NSA[b]
1	Austin and Pickersgill[c] Sunderland £6 669 000 (Shipbuilders; nationalized)	Sunderland Shipbuilders[c] Sunderland £15 000 000 (Shipbuilders; nationalized)	T. I. Churchill Blaydon £776 000 (Machine tools; subsidiary of T. I. Group)	Sunderland Shipbuilders[c] Sunderland £17 005 000 (Shipbuilders; nationalized)
2	Lombard North Central Leasing—[d] £6 453 000 (Leasing Company; subsidiary of Nat.West Bank)	Austin and Pickersgill[c] Sunderland £9 000 000 (Shipbuilders; nationalized)	George Blair[c] Newcastle £620 700 (Steel foundry; local independent company)	Austin and Pickersgill Sunderland £15,669,000 (Shipbuilders; nationalized)
3	British Steel Corporation Consett £5 322 000 (Steelworks; nationalized—closed 1980)	NEI Parsons[c] Heaton £6 723 000 (Electrical and nuclear engineering; HQ in Newcastle)	Welwyn Electric Bedlington £583 528 (Electronic components; subsidiary of Royal Worcs.)	NEI Parsons[c] Heaton £6 976 401 (Electrical and nuclear engineering; HQ in Newcastle)
4	Federation Brewery[c] Gateshead £4 838 000 (Brewery; HQ in Newcastle)	Brentford Nylons Cramlington £5 301 000 (Household textiles; subsidiary of Lonrho; HQ in London)	E. Jopling and Sons[c] Sunderland £445 050 (Steel foundry; HQ in Glasgow)	Lombard North Central Leasing—[d] £6 453 000 (Leasing company; subsidiary of Nat.West Bank)
5	Scottish and Newcastle Brewery[c] Newcastle £4 249 000 (Brewery; HQ in Edinburgh)	Sterling Winthrop[c] Newcastle £4 220 000 (Pharmaceuticals; HQ in New York)	Bonas Machine[c] Sunderland £369 478 (Textile Machinery; local independent company	Sterling Winthrop[c] Newcastle £5 658 000 (Pharmaceuticals; HQ in New York)
6	Alcan Aluminium Ashington £4 133 000 (Aluminium smelting; subsidiary of Canadian company)	Boots Cramlington £2 500 000 (Pharmaceuticals; HQ in Nottingham)	Davy Roll[c] Gateshead £277 462 (Steel foundry; HQ in London)	Boots Cramlington £5 444 000 (Pharmaceuticals; HQ in Nottingham)
7	National Smokeless Fuels[c] Jarrow £3 621 000 (Coke works; subsidiary of NCB)	Coles Cranes[c] Sunderland £2 443 600 (Manufacture of cranes and lifting gear; HQ in London)	NEI Parsons[c] Heaton £253 401 (Electrical and nuclear engineering; HQ in Newcastle)	British Steel Corporation Consett £5 322 000 (Steelworks; nationalized—closed 1980)

8	NEI[c, e] Newcastle £3 403 000 (Multi-plant engineering company; HQ in Newcastle)	Plessey[c] South Shields £1 900 000 (Manufacture of printed circuit boards; HQ in Essex)	British Engines[c] Newcastle £230 632 (Precision engineering; local independent company)	Brentford Nylons Cramlington £5 301 428 (Household textiles; subsidiary of Lonrho; HQ in London)
9	Boots Cramlington £2 944 000 (Pharmaceuticals; HQ in Nottingham)	NEI, Clarke Chapman[c] Gateshead £1 777 560 (Marine engineering; HQ in Newcastle)	Whitley Bay Meat Supply Whitley Bay £181 819 (Wholesale meat suppliers; local independent company)	Fédération Brewery[c] Gateshead £5 838 000 (Brewery; HQ in Newcastle)
10	Vaux Breweries[c] Sunderland £2 655 000 (Brewery; HQ in Sunderland)	Vickers[c] Newcastle £1 597 320 (Armaments manufacture; HQ in London)	Victor Products Wallsend £138 460 (Mining equipment; local independent company)	Scottish and Newcastle Brewery[c] Newcastle £4 262 000 (Brewery; HQ in Edinburgh)
Total	£44 287 000	£50 462 908	£3 876 530	£76 928 829
% of known expenditure	26.7	58.5	70.6	29.8

[a] For each establishment, the name, location, amount of assistance, and activity and ownership are given (the details of activity and ownership relate to the position around 1983–4.

[b] The table includes only published amounts; in the case of RDG these are payments greater than £25 000, and in the case of RSA and NSA approvals above £5000. However, it is estimated that published amounts account for 79 per cent of total expenditure on these three schemes.

[c] Establishment located in designated inner area.

[d] The ultimate beneficiaries of RDG claimed by Lombard North Central Leasing are not known, and so payments made to this company could not be ascribed.

[e] Because of the manner in which the figures were published, this expenditure could not be allocated to any particular establishment of NEI.

Source: DTI.

Table 4.10. *Employment Size of Financially Assisted Establishments*[a] *in the Newcastle Metropolitan Region, by Agency/Scheme, 1974–1984*

Agency/Scheme	Size group (%)			Average employ-ment	Median size group
	1–25	26–200	>200		
DTI—Of which:	11.7	24.2	64.1	639	201–500
RDG	(7.2)	(18.8)	(74.0)	(760)	(201–500)
RSA	(27.6)	(45.7)	(26.7)	(237)	(51–100)
NSA	(35.8)	(32.1)	(32.1)	(199)	(51–100)
Local authorities—Of which:	79.0	19.0	2.0	27	6–10
District councils	(84.6)	(14.2)	(1.2)	(19)	(6–10)
County councils	(61.4)	(34.3)	(4.3)	(52)	(11–25)
English Estates	68.2	25.8	6.0	52	11–25
BTG	36.4	54.5	9.1	75	25–50
Washington Develop Corp.	90.0	10.0	0	13	6–10
BSC(I)	87.7	12.3	0	12	1–5
UDG	NK[b]	NK[b]	NK[b]	NK[b]	NK[b]
Enterprize zone	NK[b]	NK[b]	NK[b]	NK[b]	NK[b]
All agencies/schemes	52.7	21.2	26.1	265	11–25

[a] An establishment which has received assistance under two or more separate approvals (or payments in the case of RDG) will occur more than once in this table. The table includes only establishments in Tyne and Wear and County Durham, as employment figures for establishments in Northumberland are generally not known.
[b] NK, not known.

Sources: British Business and relevant agencies.

establishments; over half of the BSC(I) assistance cases, for example, involved establishments of 1–5 employees.

It is not possible to compare the employment size distribution of assisted establishments with the size distribution of *all* establishments in the area, but it is possible to make a comparison within the manufacturing sector. 61.2 per cent of manufacturing establishments in the area in 1978 were in the size range 1–25 employees. However, over the period 1974–84, only 52.7 per cent of assistance cases were in this size range. Thus the assistance system as a whole does not seem (at the least) to have been biased in favour of the small-firms sector. This is supported by Table 4.11 which indicates the proportion of total expenditure going to the top 10 per cent of assisted establishments. While the concentration of DTI expenditure is to be expected in the light of findings already reported, the concentration of local authority assistance is perhaps more surprising, with 49.3 per cent of assistance being awarded to the top decile of aided establishments.

The size of assisted establishments is related to their ownership, with smaller establishments more likely to be locally owned. We have identified a high degree of external control of industry as characteristic of the local economy. Unfortunately, comprehensive data on the ownership of assisted estab-

Table 4.11. *Financial Assistance Going to the Most Heavily Assisted Establishments in the Newcastle Metropolitan Region, by Agency/Scheme, 1974–1984*[a]

Agency/Scheme	Known expenditure going to the 10% most heavily assisted establishments by each agency/scheme (£000)	Total known expenditure (£000)	% of total known expenditure devoted to the 10% most heavily assisted establishments
DTI—Of which:	180 247	257 841	69.9[b]
RDG	(98 077)	(166 098)	(59.0)[b]
RSA	(67 128)	(86 255)	(77.8)
NSA	(3 372)	(5 488)	(61.4)
Local authorities—Of which:	7 201	14 596	49.3
in Tyne and Wear	(6 216)	(13 444)	(46.2)
outside Tyne and Wear	(887)	(1 152)	(77.0)
English Estates	3 319	6 374	52.1
BTG	1 462	5 615	26.0
Washington Development Corp.	929	2 416	38.5
BSC (I)	645	1 446	44.6
All agencies/schemes	242 796	288 288	84.2[b]

[a] Data for UDG and Enterprise Zone, assistance not available.
[b] These figures exclude small RDG payments.

Source: British Business and relevant agencies.

lishments compared with all establishments in the region are not yet available. However, the survey of assisted establishments (which includes establishments assisted since 1980) indicated that 79.0 per cent of local authority-assisted *cases* were local independent businesses. Surprisingly, this figure is not much higher than for DTI assistance cases, of which 68.3 per cent were locally owned companies. The degree of orientation towards the indigenous sector therefore does not appear to differ significantly between local authority and DTI assistance in terms of the proportion of *cases*, although this must be viewed in the context of the concentration of DTI *expenditure* on large enterprises subject to a high degree of external control.

Employment effects

Clearly, the major problem of the local economy is unemployment. To what extent has financial assistance resulted in the creation of new employment opportunities, particularly for the most deprived residents, including those living in the Department of the Environment designated inner area? The surveys of employers and employees shed some light on these questions.

Table 4.12 reveals that in a majority of cases projects which received

Table 4.12. *Effects of Assisted Projects in the Newcastle Metropolitan Region upon Employment: Increased or Decreased Employment*

Effect on employment	All projects		RDG-assisted projects[a]		RSA/NSA-assisted projects[a]		Local authority-assisted projects[a]	
	(No.)	(%)	(No.)	(%)	(No.)	(%)	(No.)	(%)
Increased employment	137	53.9	74	57.4	39	75.0	75	57.7
Retained employment	37	14.6	29	22.5	7	13.5	10	7.7
Reduced employment	5	2.0	4	3.1	2	3.8	0	0.0
No effect	58	22.8	19	14.7	3	5.8	35	26.9
Not stated/unknown	17	6.7	3	2.3	1	1.9	10	7.7
Total projects	254	100.0	129	100.0	52	100.0	130	100.0

[a] These figures refer to the projects assisted from each of the three named sources. In addition, these projects may also have had assistance from other sources (for example, RDG-assisted projects include some projects which received RSA/NSA as well).

Source: Survey of (201) assisted establishments, 1985.

assistance were reported to have led to some increase in employment. RSA/NSA aid was the most successful in this respect, with 75.0 per cent of the projects assisted leading to an increase in employment, but surprisingly there was little difference between RDG-assisted projects (57.4 per cent) and those assisted by the local authorities (57.7 per cent). Significantly more of the managers of establishments in the latter group stated that assistance had had *no* effect on employment. The RDG-supported projects more often led to the retention of employment. These contrasts could be due to the very different scale of local authority and DTI assistance.

Financial assistance to the 201 surveyed establishments was reported to have resulted in the creation of 1118 new jobs: 38.7 per cent of the jobs were for skilled and 28.6 per cent for semi-skilled workers. A high proportion (71.0 per cent) of these vacancies were reported to have been filled by people who were previously unemployed—a reflection of the high level of unemployment present in the metropolitan region (see Table 4.13). However, second-round effects have not been investigated; the characteristics of those who filled jobs vacated by the previously employed are, therefore, not known.

To what extent are the extra jobs created as a result of the implementation of assisted projects filled by residents of the inner areas, particularly those disadvantaged in the labour market? In short, who gets the jobs?

To address these questions, a survey of employees at the assisted establishments was carried out, and, drawing on this survey, an attempt was made to look at the characteristics of employees most likely to have been affected by assisted projects; this was done by identifying those recruited subsequent to the provision of assistance to an establishment. These 'post-project'

Table 4.13. *Effects of Assisted Projects in the Newcastle Metropolitan Region upon Employment: Profile of Additional Jobs*

Outcome of project	No. of additional jobs	% of total additional jobs
Total additional male employment	715	64.0
Total additional female employment	403	36.0
Total additional employment	1118	100.0
Of which:		
Jobs filled by those previously unemployed (registered or unregistered)	794	71.0
Types of jobs/occupations:		
Extra managerial jobs	112	10.0
Extra clerical jobs	69	6.2
Extra technical jobs	70	6.3
Extra skilled jobs	433	38.7
Extra semi-skilled jobs	320	28.6
Extra unskilled jobs	62	5.5
Extra apprentices	25	2.2
Not stated	27	2.4

Source: Survey of (201) assisted estalishments, 1985.

employees represent an approximation to the beneficiaries of assisted projects. Table 4.14 provides a comparison between these post-project employees and all the surveyed employees at the assisted establishments. Again, here we have only been able to consider 'first-round' employment effects.

The employees' survey revealed that 13.5 per cent of *all* sampled employees at assisted establishments lived in the designated inner area, compared with 14.7 per cent of those recruited 'post project'. Hence, the impact of assisted projects on the spatial labour market appears to simply reflect the existing spatial distribution of employees: post-project recruitment seems to follow the established patterns in the labour market.

Table 4.14 shows that a large proportion—56 out of 97 (57.7 per cent)—of the inner area residents in the post-project sample worked at inner area establishments; this data, together with Census information, does point to the high dependence of inner area residents on inner area jobs. However, it is important to stress that many inner area jobs are filled by commuters from elsewhere: 62.9 per cent of employees recruited post project at inner area establishments came from the outer areas of the Newcastle metropolitan region; however, the vast majority (90.1 per cent) were residents of Tyne and Wear, and so these employment benefits were virtually all contained within the county. Assisting inner area establishments does not, therefore, necessarily result in inner area residents being the prime beneficiaries of newly created employment. In addition, the analysis of the skill characteristics of post-project employees indicates that a large proportion of jobs created at

Table 4.14. *Characteristics of Employees in Assisted Establishments in the Newcastle Metropolitan Region, by Inner and Outer Areas and by Occupational Group*

Characteristic	All employees		Employees recruited post project							
			All establish-ments		Inner area[a] establishments		Outer area[b] establishments		(Tyne and Wear establish-ments)	
	(No.)	(%)	(No.)	(%)	(No.)	(%)	(No.)	(%)	(No.)	(%)
Place of residence:										
Inner area[a]	188	13.5	97	14.7	56	37.1	41	8.1	87	29.4
Outer area[b]	1 200	86.5	561	85.3	95	62.9	466	91.9	209	70.6
(Tyne and Wear)	(559)	(40.3)	(308)	(46.8)	(136)	(90.1)	(172)	(33.9)	(268)	(90.5)
Total	1 388	100.0	658	100.0	151	100.0	507	100.0	296	100.0
Occupational group:										
Managerial/ technical	263	18.9	105	16.0	13	8.6	92	18.1	37	12.5
Skilled non-manual	243	17.5	113	17.2	39	25.8	74	14.6	76	25.7
Skilled manual	471	33.9	224	34.0	73	48.3	151	29.8	109	36.8
Semi-skilled and unskilled manual	391	28.2	207	31.4	26	17.2	181	35.7	71	24.0
Not stated	20	1.5	9	1.4	0	0.0	9	1.8	3	1.0
Total	1 388	100.0	658	100.0	151	100.0	507	100.0	296	100.0

[a] 'Inner area' as defined by the Department of the Environment.
[b] 'Outer area' refers to the rest of the Newcastle metropolitan region—that is the study area excluding the Department of the Environment inner area.

Source: Survey of employees of (120) assisted establishments, 1985.

inner area establishments may be relatively skilled and not especially geared towards meeting the needs of deprived residents. 82.7 per cent of the post-project jobs at the inner area establishments were managerial/technical or skilled.

This analysis does indicate that there are difficulties with the approach taken in the Urban Programme through which inner area firms are assisted with the intention of benefiting deprived inner area residents. If the benefits are distributed through the 'normal' functioning of the labour market, benefits will be dispersed through commuting: in the case of the Newcastle metropolitan region, this dispersal appears to be substantial, although little of it extends beyond the county.

The cost-effectiveness of assistance

One way of assessing the cost-effectiveness of public sector assistance in generating employment is to calculate the 'cost per job': this is an exceedingly difficult task, fraught with theoretical and practical problems. Here we can do no more than briefly explore the issue, indicating the relative magnitude of costs in relation to jobs created through the implementation of assisted projects.

In the employers' survey, managers were asked about the numbers of new jobs associated with projects which had received assistance, and gave details of the amounts and sources of assistance. From this it becomes possible to calculate a crude cost-per-job figure for all assistance, including those projects which did not lead to the creation of new jobs. However, this approach is of limited utility and it is possible to make some refinements. In particular, it is useful to take into account whether the assistance was 'additional'—that is, actually necessary to secure the implementation of a project—or represented 'deadweight' spending. In the employers' survey, managers were asked to comment on the likely outcome if individual projects had not been assisted, and this provided a way of identifying 'additional' and 'deadweight' spending[2] . Because some projects were multi-funded, it was necessary to consider the outcome in relation to each individual source of assistance, and the analysis given below is based on this approach.

It should be noted at the outset that such an analysis of cost-effectiveness, even though additionality is incorporated, is based on a fairly restricted definition of cost per job. The survey excluded establishments which had received assistance and failed to survive. The figures take no account of displacement effects (that is, where jobs created in an assisted establishment lead to lost business and lost jobs in another establishment), nor do they take into account the multiplier effects of assistance within the economy. Moreover, the analysis is focused on comparisons between existing policy measures and does not consider alternative ways in which the resources could have been used for job creation; it does not introduce the broader costs and benefits (such as savings to the Exchequer of reduced unemployment costs, or revenues in the form of taxes); and, finally, the approach is quantitative (the number of jobs) rather than qualitative (skills, gender, conditions, etc.).

Table 4.15 shows cost-per-job calculations in relation to projects assisted from each of the three main sources (RDG, RSA/NSA, and the local authorities). It should be noted that each category, by source, covers all projects aided by that source—including those also receiving aid from other sources. A number of different measures are given in an attempt to deal with the problem of assessing cost per job when projects are multi-funded.

The first column of Table 4.15 shows the total number of new jobs associated with the projects assisted from each of the three sources. In the case of RDG, 593 jobs were associated with 122 projects (column 2) involving £5.4 million of RDG assistance (column 3). Some of these projects were aided from other sources: hence, *total* public sector support to the RDG 'client group' amounted to £8.3 million (column 4). The crude cost per job in the case of

Table 4.15. *'Cost per Job' for Assisted Projects in the Newcastle Metropolitan Region*

Source of assistance	'Crude' cost per job					'Refined' cost per job				
						Projects where assistance from named source was 'additional'		Projects where *only* the assistance from the named source was 'additional'		
	1. No. of new jobs associated with projects assisted by named source	2. No. of projects	3. Total assistance paid to projects from named source (£)	4. Total assistance paid to projects (all sources) (£)	5. 'Crude' cost per job (£) (column 3 / column 1)	6. No. of new jobs	7. 'Refined' cost per job (£) (column 3 / column 6)	8. No. of new jobs	9. Total assistance paid by named source (£)	10. 'Refined cost per job (£) (column 9 / column 8)
RDG	593	122	5 421 000	8 287 000	9 142	307	17 658	72	657 000	9 125
RSA/NSA	522	48	2 278 000	6 614 000	4 364	317	7 186	112	399 000	3 563
Local authority	567	123	1 366 000	3 459 000	2 400	157	8 701	93	201 000	2 161

Notes: 1. All assistance figures are cost prices.
2. Projects receiving assistance from named source (RDG; RSA/NSA; local authorities) may also have been assisted from other sources.
3. In columns 6–10, projects where assistance was 'additional' refers to those projects which managers said would have been delayed, cancelled, or have taken place elsewhere had assistance not been given.

Source: Survey of (201) assisted establishments, 1985.

RDG, calculated by dividing the total RDG assistance paid by the total number of new jobs was nearly £9150. However, this is a crude measure because it includes 'deadweight' spending (where the RDG was not necessary to secure the implementation of the project) and also because some of the new jobs may have stemmed from assistance from other sources; of the total of £8.3 million paid to the RDG 'client group', 36 per cent came from sources other than RDG.

In the case of the RSA/NSA projects, 522 new jobs were associated with projects receiving altogether £2.3 million of RSA/NSA, giving a crude cost per job of over £4350—about half the comparable figure for RDG. The local authority crude cost-per-job figure was well below this, at £2400. However, both the RSA/NSA-assisted and local-authority-assisted projects attracted substantial proportions of assistance from other sources—much higher than was the case for the RDG group.

Columns 6–10 of Table 4.15 refer to projects where assistance was 'additional'. Column 6 shows the number of new jobs associated with those projects where the stated source of aid was 'additional'; thus 307 of the 593 jobs associated with RDG-assisted projects were created in projects which were said to have depended on the provision of RDG assistance. On this basis, 'refined' cost-per-job figures were calculated and are shown in column 7. Because 'non-additional' jobs are excluded, these cost-per-job figures are higher than the crude figures (in column 5). RDG is again shown to have a relatively high cost-per-job figure. The difference between RSA/NSA and local authority costs are reduced and reversed because of the large proportion of discounted non-additional jobs in local authority projects. However, this has to be treated with caution since these 'client groups' may still have had assistance which was 'additional' from other sources.

The final part of the table includes only those projects where the named source of assistance was 'additional' and, thus, here it is possible to attribute jobs more directly to the source of assistance; in the main, these are single-funded projects. Once again, the cost-per-job figure for RDG projects is relatively high (£9100) in comparison with the figures for projects assisted by RSA/NSA (£3550) and local-authority-aided projects (£2150).

The overall finding from the figures produced in this analysis is that the cost per job in RDG-assisted projects appears to be significantly higher than that for RSA/NSA-aided and local-authority-aided projects; this contrast is shown throughout the analysis using a variety of methods (although the earlier caveats must be borne in mind). This may well be related to the concentration of RDG on large capital-intensive manufacturing industries, and it would therefore seem that automatic capital grants may not be the most cost-effective way of creating new jobs. On the other hand, there may well be good grounds for supporting large capital-intensive industries which are the mainstays of the area's economy and upon which many other local businesses depend.

The cost-per-job figures do, however, point to the benefits of a more selective approach—such as RSA/NSA and, to a large degree, local authority assistance. These assistance schemes have a lower cost per job than RDG. Moreover, in the case of RSA, assistance is actually linked directly to job creation or retention, and this is often a criterion for local authority assistance. This may, therefore, be regarded as providing some justification for the recent change in DTI policy which has made the provision of RDG more dependent upon job creation—a recognition of the need to introduce greater selectivity to improve cost-effectiveness. Furthermore, as we argue below, selectivity is an essential component of a strategic, targeted, and effective approach to economic development policy.

Increasing the effectiveness of policy

In reviewing econonomic development it is necessary, at the outset, to stress the scale and nature of the problems which policy aims to address: long-term economic decline, deep-seated structural problems, and high levels of unemployment. The resources employed to tackle these issues, while substantial, have in fact been modest in relation to the scale of the problems: hence, the impact of policy (sustained over a long period) has not been sufficient to resolve the economic difficulties of the Newcastle metropolitan region. Moreover, it must be borne in mind that policy has been limited in its application to the types of problems we have identified and that, in addition, the period examined in this study has included severe recession which has exacerbated existing problems and made policy intervention more difficult.

However, urban and regional policies have had some important impacts and the resources involved have, by no means, been trivial. Could *greater* beneficial impacts have been derived had these resources been distributed *differently*? What scope is there for increasing the effectiveness of policy?

Our analysis of financial assistance suggests that there is scope for enhancing the effectiveness, and increasing the impact, of policy. A large proportion of assistance has been provided on a *reactive*, rather than *pro-active*, basis; in the case of RDG in particular, assistance has responded to investment rather than being used to lead and direct economic change. In our view, there is a need for a pro-active approach based on clear objectives and explicit targeting on particular sectors, sizes of firms, locations, and specific parts of the labour market. There is need for greater selectivity so as to target assistance in relation to perceived problems and objectives, and also to ensure that assistance is provided where it is most needed. The survey indicated a large element of deadweight spending which, we believe, could be reduced through a more selective approach: the 1984 revisions to RDG have only gone some way towards this by reducing the concentration of RDG on capital-intensive projects. In addition, cost-per-job calculations, while tentative and necessarily hedged with caveats, do indicate that selective assistance is significantly more cost-effective than automatic investment grants in generating employment. The need for greater selectively seems also to be evident in relation to urban policy: for example, some selective schemes (such as wage subsidies) may be more efficient ways of targeting aid to inner area residents than the provision of more general forms of assistance to inner area establishments.

In considering how to improve the effectiveness of assistance policy, it is necessary to examine the institutional framework. One of the most striking features of the current assistance system is the large number of agencies and schemes: the DTI (RDG, RSA, NSA); thirteen out of the fourteen local authorities in the study area which provide financial assistance; English Estates and Washington Development Corporation (rent relief); BSC(I) (loans and grants); UDG and Enterprise Zone rate relief. This list could be considerably extended by the addition of agencies offering other forms of assistance (for example, advisory services) and carrying out related initiatives (for example, the MSC). In many ways this multiplicity of agencies is actually a strength rather than a weakness. There is not a *major* duplication of effort: each of the agencies has particular objectives, and aims to serve particular 'clients'. Nor does there seem to be very substantial overlap: over the period 1974–84, 83.6 per cent of assisted establishments received assistance from just one source, although this does not take into account the smaller RDG payments (Table 4.16). The various agencies and schemes appear, in general, to be complementary rather than in conflict, and often well matched to the 'markets' they serve and the kinds of assistance they offer. Thus the DTI has predominantly been concerned with large establishments and large projects while the local authorities have been well placed to become informed about small firms in their areas and to assist them. Likewise, those agencies (including the local authorities) which provide premises are usually best able to administer rent relief.

However, the strengths of an assistance system operating through a variety of specialist agencies may be countered by weaknesses—problems and inefficiencies—resulting from a lack of co-ordination. There is no coherent framework for action and therefore no concerted attempt at strategic targeting to address the problems of the metropolitan region as a whole. At present, each of the agencies makes its contribution in relation to its own remit and perception of the problems rather than acting together with the other agencies to pursue an overall and integrated strategy: there is no agreed strategy to concentrate agencies' resources on, for example, specific sectors, areas, or groups within the labour market.

The first requirement is to establish an agreed strategy, involving a clear identification of the problems of the metropolitan region and the kinds of policies (and not only financial assistance measures) needed to address these issues. Conflicts between long-term structural policies (for example, aiding particular growth sectors) and more immediate responses (for example, targeting aid to support specific groups in the labour market) would need to be

Table 4.16. *Overlap of Financial Assistance Activities between Agencies/Schemes in the Newcastle Metropolitan Region, 1974–1984*

Agency/scheme	No. of establishments receiving assistance from named agency/scheme and:						Total no. of establishments assisted by agency/scheme	% of establishments receiving aid only from named agency/scheme
	No other	1 other	2 others	3 others	4 others	5 others		
DTI:								
RSA	141	125	69	15	0	2	352	40.1
NSA	21	28	17	5	0	1	72	29.2
RDG	251	105	44	10	0	2	412	60.9
Tyne and Wear district councils	955	99	30	5	0	1	1090	87.6
Tyne and Wear County Council	124	49	18	9	0	0	200	62.0
Durham district councils	29	4	1	1	0	0	35	82.8
Durham County Council	45	27	21	6	0	1	100	45.0
Northumberland District councils	7	2	2	0	0	0	11	63.6
Northumberland County Council	34	2	0	0	0	0	36	94.4
English Estates	184	79	47	12	0	2	324	56.7
BTG	15	4	4	1	0	2	26	57.7
Washington Development Corp.	172	31	9	1	0	0	213	80.8
BSC(I)	74	31	14	7	0	1	127	58.3
Total no. of cases	2052	586	276	72	0	12	—	—
Total no. of establishments	2052	293	92	18	0	2	2457	—
% of total firms	(83.6)	(11.9)	(3.7)	(0.7)	(0)	(0.1)	(100)	—

Notes: The table includes only those payments for which details are known. An establishment may have been assisted by the same agency/scheme more than once over the period 1974–84, but this is included as just one case. Reading across the table gives the number of establishments assisted by each named agency/scheme. However, care is needed when reading down the columns as there is an element of double-counting: for example, there are 586 cases where an establishment has received assistance from two agencies/schemes over the period, but since each establishment is included twice in this column, there is therefore only half that number of actual establishments—that is, 293.

Sources: *British Business* and relevant agencies.

resolved. It would evidently be necessary to establish the relative priority (and, hence, expenditure) to be accorded to helping areas of especially high unemployment (for example, the inner areas) as contrasted with an approach to target aid at the unemployed throughout the area. Moreover, the development of a strategy would highlight the need and provide the explicit capability of making trade-offs in terms of the distribution of scarce resources of finance and expertise.

The recent establishment of City Action Teams (CATs) represents an attempt to improve the co-ordination of policy. The CATs consist of the Regional Directors of the Department of the Environment, the DTI, and the Department of Employment and the MSC, and the intention is that they will provide an opportunity for these departments to co-ordinate their activities and ensure that action is well related to local needs and opportunities. It is intended that the CATs will develop priorities relevant to local needs, establish joint working arrangements, and monitor the output of programmes and projects. A CAT has been set up to operate within the Newcastle–Gateshead Inner City Partnership Area to perform these functions.

While this initiative is to be welcomed, it does not go far enough towards meeting the need for a co-ordinated strategy. Firstly, it does not embrace all the agencies involved in economic development. Secondly, it only covers the Partnership area (it does not even include the Programme areas); and, thirdly, it seems essentially to provide an opportunity to co-ordinate existing policy approaches rather than generate new, perhaps more relevant or targeted, policies.

In our view, the CAT approach needs to be built on and extended. There is a need to develop a strategy covering the whole metropolitan region in view of the labour market interdependences existing within this functional region. It should, to be effective, include all the main agencies involved in economic development, making them part of a coherent framework, rather than be organized around departmental/agency concerns. The activities of each agency should be tailored to the needs identified in the strategy and reflect their specialisms and capacities. In some cases, there will be a need to increase capacities (for example, to expand administration to operate more selective policies) and, in other cases, opportunities to share expertise.

Finally, it is vital that such a strategy combines the concerns of both 'urban' and 'regional' policy; it must also recognize the need to pursue both industrial and social policy initiatives to respond to the continuing and severe problems of the Newcastle metropolitan region.

Notes

1. The inner area shown in Figure 4.3 does not perfectly match the Department of the Environment designated inner area because some wards straddle the inner area boundary; it is, however, a close approximation.

2. Those projects where spending was 'additional' would, according to managers, have been delayed, have been cancelled, or have taken place elsewhere in the absence of assistance while 'deadweight' spending refers to assistance for projects which would have gone ahead even without assistance.

References

Coombes, M.G., Dixon, J.S., Goddard, J.B., Taylor, P.J., and Openshaw, S. (1980), 'Functional Regions for the 1981 Census of Britain: A User's Guide to the CURDS Definition', *CURDS Discussion Paper* 30, Centre for Urban and Regional Development Studies, Newcastle upon Tyne.

—— —— —— —— (1981), 'Appropriate Areas for Census Analysis: An Outline of Functional Regions', *CURDS Discussion Paper* 41, Centre for Urban and Regional Development Studies, Newcastle upon Tyne.

Department of the Environment (1983), *Urban Development Grant: Guidance Notes*, London: Department of the Environment.

—— (1985), *Urban Programme: Ministerial Guidelines*, London: Department of the Environment.

Ganguly, P. (1984), 'Business Starts and Stops: Regional Analysis by Turnover, Size and Sector, 1980–83, *British Business*, 18 May, 10–13.

Goddard J.B, (1983), 'The Economic Development of Older Industrial Areas: The Case of the Northern Region of England', *CURDS Discussion Paper* 53, Centre for Urban and Regional Development Studies, Newcastle upon Tyne.

HMSO (1963), *The North-East—A Programme for Regional Development and Growth* (Hailsham Report), Cmnd. 2206, HMSO.

House, J.W. (1969), *Industrial Britain: The North-East*, Newton Abbot: David and Charles.

McCord, N. (1979), *North-East England: The Region's Development 1760–1960*, London: Batsford.

Manners G., Keeble, D., Rodgers, B. and Warren, K. (1972), *Regional Development in Britain*, London: Wiley.

Mess, H.A. (1928), *Industrial Tyneside: A Social Survey, London: Ernest Benn.*

NECCA (1984), *The State of the Regional Report 1984*, Newcastle upon Tyne: North of England County Councils Association.

Northern Region Strategy Team (1977), *Strategic Plan for the Northern Region*, Vol. 2: *Economic Development Policies*, London: HMSO.

Odber, A.J. (1965), 'Regional Policy in Great Britain', in US Department of Commerce, *Area Development Policies in Britain and the Countries of the Common Market.*

Pepler, G. and MacFarlane, P.W. (1949), *North-East Area Development Plan*, London: Ministry of Town and Country Planning.

Sinfield, A. and Fraser, N. (1985), *The Real Cost of Unemployment*, Newcastle: BBC North-East.

Storey, D.J. (1982), *Entrepreneurship and the Small Firm*, London: Croom Helm.

—— (1985), 'Manufacturing Employment Change in Northern England 1965–78: The Role of Smaller Businesses', in D. J. Storey (ed.), *Small Firms in Regional Economic Development; Britain, Ireland and the United States*, Cambridge University Press.

Tyne and Wear County Council (1982), *Manufacturing Employment Change in Tyne and Wear since 1965*, Tyne and Wear County Council, Department of Planning.

—— (1985), *Unemployment in Tyne and Wear*, Tyne and Wear County Council, Department of Planning.

Whittington, R. (1984), 'Regional Bias in New Firm Formation', *Regional Studies*, 18, 3.

5

The Inner Cities Research Programme: The Clydeside Case-Study

*W. F. Lever**

Introduction

Within the Economic and Social Research Council (ESRC) Inner Cities Research Programme, the Clydeside conurbation was selected as an example of a poorly adapting local labour market with a long-run history of economic decline as a result of which unemployment rates are currently, and have been historically, high relative to national figures. The economy of Clydeside has changed radically in the post-war period as the old manufacturing base of iron and steel, heavy engineering, and shipbuilding has declined; as new lighter industries have been established, often in locations different to those of the old industrial base; and as service sector employment has grown very substantially. At the same time, the urban fabric has undergone massive change, and the administrative structure changed greatly in the mid-1970s. Clydeside, because it has experienced worse economic and environmental problems than any other city in the United Kingdom (UK)—except possibly Belfast—has attracted the highest levels of public sector intervention in Britain. It has always, throughout the post-war period, qualified for Special Development Area status; it has always qualified for 'Urban Programme' types of assistance; and, since 1975, the Scottish Development Agency (SDA) has devoted a major share of its resources to the conurbation. Other national programmes covering Enterprise Zones, enterprise trusts, community business, various forms of labour and management training, and schemes of assistance for new business creations have all found a part to play in the economic adaptation of Clydeside. Within the Inner Cities Research Programme, therefore, Clydeside can be viewed as the example of full 'policy-on' experience. At the same time, the multiplicity of policies and agencies have made it difficult to define the results of specific programmes where enterprises may be receiving several forms of assistance simultaneously, and the uniqueness of the Scottish dimension, not least the

* This chapter summarizes the work of the Clydeside study team within the ESRC Inner Cities Research Programme. The team comprised Simon Booth, Chris Jolliffe, William Lever, Frank Mather, James McEldowney, Alan McGregor, John Money, Chris Moore, Douglas Pitt, Isobel Robertson, and Vivienne Skinner.

existence and operation of the SDA, make comparability with other depressed British cities difficult. Nevertheless, within the ESRC Programme, each study adopted broadly the same research agenda with four elements. These were: firstly, an analysis of the nature of business change in order to identify key elements in the restructuring of the local economy as a basis for policy formulation; secondly, an analysis of labour market adjustment; thirdly, an assessment of the efficacy of policies; fourthly, an examination of the objectives and functioning of the agencies with the responsibility for implementing these policies.

Business Change

Three quite clear trends in the economy of the whole of the Clydeside conurbation can be seen in the period since 1950, when measured in terms of employment. Firstly, between 1951 and 1984 there was a substantial decline in total employment. Secondly, there was a very significant shift out of manufacturing and into service activities. Thirdly, there was at least up to 1978, a considerable suburbanization of employment, leaving the city of Glasgow with a smaller proportion of the stock of jobs and leaving the surrounding ring of towns with an increasing proportion.

In summary, total employment in the whole of the West Central Scotland conurbation fell from 844 000 in 1952 to 686 000 in 1981. By 1984, it was estimated that total employment in the conurbation had fallen further to 640 000. The rate of loss steepened from a negligible −400 per year in the 1950s to −4700 per year in the 1960s and early to mid-1970s, to −25 000 per year in the late 1970s and early 1980s. The rate of decline in the inner city, which we define as the administrative city of Glasgow, has been continuous; the outer city, which comprises the rest of the built-up area, and includes old industrial towns, two New Towns, and suburban dormitories, grew in employment numbers until the early 1970s and then declined. One intriguing element in the pattern of change is that, while the inner city's share of total employment fell from 66.8 per cent in 1952 to 57.2 per cent in 1978, there was the beginning of an indication of a relative recovery as it rose to 58.2 per cent by 1981. Estimates of future employment change to 1987 indicate that this trend will continue and that the figure for 1987 might be 58.5 per cent.

Rather different patterns of change occurred in the service and manufacturing sectors after 1961. For the conurbation as a whole, employment in manufacturing fell from 387 000 in 1961 to 187 544 in 1981, with faster rates of loss in the 1960s (−8000 per year) and in the late 1970s (−23 000 per year) than in the early 1970s. If the more generously defined 'industries of production' (which include construction and public utilities) are considered, then employment fell from 482 000 to 252 000. Service employment grew from 350 000 in 1961 to 407 000 in 1971 and to 431 000 in 1978, since when it has remained fairly stable. In manufacturing, the 1960s were a period of dramatic job loss in the inner cities, while there was growth in the outer ring; since 1971 the rate of manufacturing job loss has been lower in the inner city and the greatest rate of loss has been in the outer city. Thus the inner city had 58 per cent of the manufacturing jobs in 1961, 49 per cent in 1971, and 52 per cent in 1981. In services, the inner city's job total rose up to 1978 and has subsequently declined by 10 000 or 4 per cent; in the outer city the growth has continued to the present, with 33 000 additional jobs between 1971 and 1981 (+24 per cent). In consequence, the inner city had 74 per cent of the service jobs in 1961, 67 per cent in 1971, and 61 per cent in 1981.

Finally, in common with most cities in the developed world, the proportion of the work-force in services has risen in the city as a whole, from 42 per cent in 1961 to 52 per cent in 1971 and 63 per cent in 1981. In the administrative city of Glasgow, the corresponding percentages were 48 per cent, 59 per cent, and 66 per cent. It is no exaggeration to say that in twenty years Glasgow (and the whole Clydeside conurbation) has gone from being an industrial city, with 60 per cent of its labour force in manufacturing, to a service centre with 60 per cent of its labour force in service occupations.

Table 5.1 gives the sectoral composition of the work-force for the conurbation as a whole for 1961, 1971, and 1981. Changes in the manner in which the data were collected mean that the 1961 and 1971 data are not strictly comparable with those for 1981, but the absolute magnitude of the changes is such that this caveat can be disregarded. In the period 1961–71, employment overall declined by 50 000 or 6.1 per cent. In absolute terms, the major job losses were experienced in mechanical engineering (−28 000), mining (−11 000), shipbuilding (−19 000), and textiles (−13 000). Only two industrial sectors—instrument engineering and electrical engineering—increased employment (by 7500). In all the service sectors except distribution there were increases, especially in professional services (predominantly health and education) and miscellaneous services (personal, leisure, and social work services) and in administration. In the period 1971–81 the rate of job loss increased overall to −13.2 per cent. Every manufacturing sector lost employment—some, such as metal manufacture (iron and steel, mainly), electrical engineering, textiles, and vehicles, by approximately a half. The only major growth in employment was to be found in financial services, professional services, and miscellaneous (personal) services. Three things are clear from Table 5.1. Firstly, the extent of the job loss is very significant and accelerating: it should be borne in mind that these figures are for the whole conurbation, peripheral suburbs included, so that none of this job loss is attributable to short-distance movement over the boundary of the study area. Secondly, while in the 1960s the manufacturing job loss was concentrated on the old industrial base, by the 1970s decline had extended to all major sectors and

Table 5.1. *Clydeside Conurbation: Employment Change, 1961–1981*

Sector	Employment 1961	% change 1961–71	Employment 1971	% change 1971–81	Employment 1981
Agriculture	3 629	−32.3	2 456	−14.9	2 090
Mining	15 037	−74.1	3 890	−6.9	3 622
Food, drink, tobacco	40 978	−8.7	37 433	−21.3	29 451
Chemicals	12 387	−25.5	9 238	−9.3	8 385
Metal manufacture	39 195	−18.5	31 948	−49.1	16 246
Mechanical engineering	86 467	−32.7	58 225	−43.2	33 068
Instrument engineering	6 158	+23.3	7 592	−24.1	5 761
Electrical engineering	18 311	+33.6	24 465	−50.2	12 180
Shipbuilding	38 416	−48.3	19 874	−35.8	12 750
Vehicles	28 035	−6.3	26 256	−45.1	14 426
Other metal goods	14 559	−25.7	10 817	−37.8	6 731
Textiles	27 016	−46.1	14 553	−47.3	7 671
Leather	2 829	−28.0	2 039	−37.3	1 277
Clothing	22 292	−15.4	18 863	−37.3	11 743
Non-ferrous minerals	9 871	−23.7	7 529	−46.8	4 004
Timber	10 486	−8.1	9 639	−32.4	6 517
Paper, printing	23 062	−7.8	21 256	−35.2	13 770
Other manufacturing	7 143	−10.6	6 388	−44.2	3 563
Construction	67 632	−12.1	59 481	−11.9	52 389
Gas, electricity, water	12 308	−13.8	10 605	−18.0	8 694
Transport	58 928	+6.4	62 687	−20.7	49 709
Distribution	113 976	−10.3	102 273	−16.0	84 929
Insurance, banking	16 302	+38.05	22 582	+20.8	27 273
Professional services	91 674	+9.6	100 516	+22.7	123 322
Miscellaneous services	55 987	+34.2	75 160	+33.9	100 648
Public administration	24 447	+78.1	43 549	+4.1	45 319
Total	840 882	−6.1	789 514	−13.2	685 538

Source: Censuses of Employment.

into services. Thirdly, in both time-periods the expanding sectors were, because of the nature of the jobs involved, unlikely to be taking in workers released by the declining sectors. The declines were greatest in the semi-skilled, male, manual groups; the increases were in the clerical and professional non-manual groups, with a much higher proportion of female workers. For the inner city the major features are the large job losses in iron and steel (metal manufacture), mechanical engineering, clothing, and paper and printing, and the failure of service employment increases to be sustained after 1978. For the outer city there were major employment losses in iron and steel (including those sustained by the large integrated Ravenscraig steelworks), mechanical engineering, electrical engineering, and the vehicle industry in 1978–81 with the closure of the Linwood car plant. Service employment, however, held up much better in the outer city after 1978.

While Clydeside's economy, in terms of employment, has clearly undergone major decline since 1961, its heavy reliance upon sectors such as iron and steel, heavy engineering, and shipbuilding which have declined nationally may merely mean that the problem is attributable solely to structural composition, rather than to locational or other factors. However, a shift–share analysis demonstrates that the structural explanation is inadequate to explain job loss. For the period 1961–71 there was national growth and the inner city had a positive structural component, whereas the outer city had a negative one. The inner city's residual component was −70 000, compared with a positive component of +30 000 for the outer city. The pattern was repeated in 1971–8 where a positive national component, a positive inner city structural component, and a negative outer city structural component generated residuals of −44 000 for the inner city but −1000 for the outer city. By 1978–81, the national component was negative, but the residuals were −18 000 for the inner city and −15 000 for the outer city. Two conclusions are clear. Firstly, the aggregate job loss for Clydeside since 1961 cannot be attributed to structural factors. Secondly, the inner city's competitive performance, as measured by the residual, is poor though the size of the job shortfall is declining slightly; however, the outer city's performance has shifted from strongly positive to strongly negative. It should be pointed out that the use of the term 'competitive' implies failure in market/private sectors, but the largest elements in computing the residual values are in the public sector services such as health, education, and social work. For example, in the period 1971–8, for the inner city the total residual figure was −44 000, but for the professional services it was −7000, for personal services it was −6000, and for public transport and communications it was −5000. These values presumably reflect, not failure in competitive performance, but declining local population requiring public sector service provision.

The analysis of employment change indicates that Glasgow, especially the inner city, has suffered an economic performance significantly worse than that of Britain as a whole on a sector-by-sector basis. The inference must be that either (1) although similar on a broadly defined product-market basis, the industrial establishments in Glasgow are different from their national comparators in terms of size, age, management, or production methodology in ways which handicap them competitively, or (2) there are aspects of Glasgow which, in terms of factor costs or other problems, make it a difficult location for industry.

The first way of approaching the nature of the problems which industry in Glasgow suffers is by what has become known as a 'components of change' analysis, which resolves employment change into its several components:

employment gains in new plant openings and *in situ* expansions, employment losses due to plant closures and *in situ* contractions, and the gross flows of jobs with plant migration.

For all three areas for which analysis was possible for 1970–80 (Scotland, Strathclyde region—the nearest equivalent to the inner and outer cities combined—and Glasgow City), there was a clear worsening in the rate of new firm formation after 1975. Not only was the rate of new firm formation lower, but the mean size of new formations was also lower. Glasgow, with new firm formation rates of 1.2 per cent per annum in 1970–5 and 0.5 per cent per annum in 1975–80, compares poorly with the rest of Strathclyde (outer city), which had corresponding rates of 2.6 per cent and 1.5 per cent per annum. The rates of growth of employment in non-new establishments were about 50 per cent higher in the outer city than in Glasgow. On contractions and closures, however, there is very little difference between Glasgow, the outer city, and the rest of Scotland in both time-periods. The crucial difference between Glasgow and the outer city, therefore, would appear to be in the rate of new firm formation, with regard to which, throughout the 1970s, Glasgow does not appear to have been a favoured location.

No comparable data exist for services, but an earlier study of the components of employment change for Glasgow for the five major service sectors (public administration is excluded) shows that relatively small net changes in employment conceal huge gross movements. In the most important sector (in terms of employment)—distribution—a net change of −7000 conceals the creation of 11 000 new jobs in 1000 new shop and warehouse openings, and a further 6000 new jobs in expansions *in situ*. However, twice as many establishments contracted, with twice the job loss, and shop closures, although only half the number of openings, had the effect of losing about an equal number of jobs (11 000).

Considerable discussion attaches to the relationship between establishment characteristics and recent employment change. Studies have demonstrated that, within manufacturing, small firms and indigenously owned firms have a better recent record of job creation or maintenance than large plants or foreign-owned plants (Mason 1986). More recently, however, some doubt has been cast upon the small-firm sector as a source of new jobs as enhanced rates of new firm creation appear to have had the effect of increasing the closure rate of small firms (Storey 1982). In the West Central Scotland conurbation, data for the period 1975–8 do indicate that the smaller establishments created employment gains whereas establishments with more than 200 employees declined by an average of more than 10 per cent in three years. The more recent time-period, 1978–81, does provide evidence that the small-firm sector has become much more vulnerable (although the data may under-record new small-firm openings). From the data used in the study, it is not possible to distinguish between single-plant locally owned firms and branch plants of UK companies, but merely between UK- and foreign-owned plants. The data do show that in 1975–8 the foreign-owned plants were still creating net additions to the employment stock of the region, whereas UK-owned plants were not. However, from sample survey data, this decline in the UK-owned sector is likely to have occurred particularly in the branch plants of non-Scottish-owned/headquartered companies rather than locally owned single-plant companies. By 1978–81 there had been a dramatic worsening of employment in foreign-owned companies where almost half of all jobs disappeared in three years.

If businesses in the inner city are more likely to fail, or less likely to expand, or if the inner city does not provide as attractive a location as elsewhere for new businesses, we would expect some evidence of this in the operating cost functions of firms in the inner city when compared with those elsewhere. There is some general evidence to show that profits per worker are lower in the conurbations than elsewhere, whether examined across all industries (Tyler *et al.* 1980) or for specific sectors (Fothergill *et al.* 1982). We have no specific data on profitability for enterprises in various parts of Clydeside or the remainder of Britain, but data from two sources give us some insight into the operating characteristics of firms in the inner city and elsewhere. Special tabulations from the 1980 Census of Production provide data on worker productivity, which is a key element in profitability (Tyler *et al.* 1984). Over the fourteen major manufacturing sectors measured in terms of gross output per head, the outer city appears to have about a 10 per cent advantage over the inner city. While such a differential might be crucial in explaining the differences in economic change, it might merely be a reflection of industrial mix—namely, that the outer city has a higher proportion of high-productivity industries such as petroleum refining, chemicals, and metal manufacture. To an extent this is true: these three sectors make up 21.8 per cent of the outer city work-force covered by the analysis but only 10.0 per cent of that of the inner city. However, in nine of the fourteen sectors, output per worker was higher in firms located in the outer conurbation than in firms in the inner city, and in some cases, such as electrical engineering and clothing, very significantly so.

The pattern of change in the economy of Clydeside is one which offers some pointers to policies which may be more effective in that they appear to be travelling in the direction of macro-economic change rather than struggling against it. The service sector will undoubtedly play a larger part in providing jobs for people. The small-firm sector, having been unfashionable in the 1960s, now appears to offer more hope for jobs than large plants of multi-plant, and often multinational, companies. The inner city areas offer more advantages to these sectors than they did in the 1960s when large manufacturing plants on green-field peripheral sites were seen as the mainsprings of economic growth and employment creation.

Table 5.2. *The Labour Force in the Clydeside Conurbation, 1971 and 1981*

Characteristics	Glasgow		Outer conurbation	
	1971	1981	1971	1981
Males				
Economically active				
Age 16–19	24 090	22 107	19 982	24 720
20–24	35 280	30 640	30 054	32 685
25–44	103 832	82 000	114 370	121 080
45–54	54 057	38 841	50 558	51 594
55+	57 405	37 490	45 208	40 998
All	274 664	211 078	260 172	271 077
Population 16–65	286 731	232 306	271 909	296 457
Economic activity rate (%)	95.8	90.0	95.7	91.4
Females				
Economically active				
Age 16–19	22 116	19 682	18 630	21 244
20–24	23 737	24 009	20 510	25 771
25–44	55 059	50 987	57 340	75 689
45–54	38 828	30 571	32 883	38 402
55+	31 891	22 128	20 386	20 674
All	171 361	147 377	149 749	181 785
Population 16–60	271 542	219 050	261 569	283 593
Economic activity rate (%)	63.1	67.3	57.3	64.1

Source: Census of Population 1971, 1981.

Labour Market Adjustment

In the last section we discussed changes in the level and structure of employment in the conurbation over time. These have obvious implications for issues such as the nature of unemployment within the conurbation. Unemployment and its composition may also reflect changes in the numbers and types of people looking for employment. The objective of this section is to examine developments in the conurbation's labour force over the same time period as the employment changes already discussed. The basic statistical evidence is presented in summary form in Table 5.2.

A number of features of Table 5.2 are worth highlighting. First, the city of Glasgow experienced a significant reduction in its labour force between 1971 and 1981. The male labour force fell by over 63 000, with a fall of 24 000 for females. The larger fall for males reflects a decline in activity rates in addition to a major drop in the population in the economically active age ranges. Secondly, the labour force in the outer conurbation registered a relatively modest growth for males, although a more significant increase for

Table 5.3. *Net Changes in Jobs and Labour Force in the Clydeside Conurbation, 1971–1981*

	Glasgow City		Outer conurbation	
	Jobs	Labour force	Jobs	Labour force
Males	−51 145	−54 425	−47 049	+10 905
Females	−11 488	−23 984	+7 387	+32 036
Total	−62 633	−88 409	−39 662	+42 941

Source: Census of Population 1971, 1981.

females. The sharp growth for females reflected an increase in the population of females, but also a jump of seven percentage points in the female activity rate for the outer conurbation. Thirdly, taking the two trends together, by 1981 well over half of the conurbation's labour force was resident in the outer conurbation, whereas the reverse position prevailed in 1971.

It is apparent from Table 5.3 that the labour force of the city of Glasgow declined by a significantly greater margin than the job loss for the decade. Conversely, in the outer conurbation the labour force grew by nearly 43 000 in net terms whereas there was a net job loss of nearly 40 000. To the extent that there is some spatial segmentation of the labour market along inner–outer lines, the position of residents of the outer conurbation should have deteriorated relative to that of Glasgow residents over the decade.

Of course, it is not entirely sensible to treat net changes in employment and labour-force aggregates as independent of each other. As the job base has declined and the probability of the unemployed finding work has fallen, some individuals may have decided to leave the labour force altogether. Thus the decline in job opportunities may bring into being a fall in the labour force. In Glasgow, the biggest proportionate fall in the male labour force was in the group aged 55 and over. There was even a decline of over 4000 in this age group in the outer conurbation. Some proportion of these net changes almost certainly reflects the discouraged-worker effect, with workers opting for earlier retirement or no longer actually seeking work because of the low probability of finding employment.

Overall, considering the conurbation as a whole, the net job loss over the decade 1971–81 far outweighed the net reduction in the labour force. Whereas a net loss of around 100 000 jobs was recorded, the labour force declined by less than 50 000 workers. One clear consequence of this is rising unemployment. The most striking feature of the behaviour of unemployment in the conurbation is quite simply the massive increase in numbers experienced over the last 10–15 years. The position is summarized in Table 5.4. Whereas there were on average around 35 000 unemployed in the conurbation in the last half of the 1960s, the 1984 average was almost 147 000.

Table 5.4. *Unemployment in the Clydeside Conurbation, 1966–1984*

	Glasgow City		Outer conurbation	
	Males	Females	Males	Females
1966–70	18 938	3 312	9 731	3 722
1971–5	27 002	4 677	16 933	6 141
1976–80	33 490	11 454	23 314	14 172
1981	50 671	18 815	41 733	23 626
1982	56 177	20 236	47 873	25 127
1983	52 883	18 933	50 575	22 361
1984	53 882	19 683	51 121	22 095

Note: The figures are annual averages.
Source: MSC.

Thus there has been a more than fourfold increase in the numbers unemployed, despite a declining labour force, changes in the method of counting the unemployed, and a proliferation of programmes designed to soak up the unemployed.

Within this aggregate growth, several other trends are apparent from an examination of Table 5.4. First, the outer conurbation has become much more important over time. In the period 1966–70, two out of three unemployed males were resident in the city of Glasgow. By 1984, the male unemployed were divided roughly equally between the city and the outer conurbation, with the majority of unemployed females resident in the outer area. In large part, of course, this simply reflects the decline in Glasgow's population relative to its outer districts.

Secondly, women have become a much more important component of unemployment over time. This is particularly pronounced for the city of Glasgow where women constituted a relatively small proportion of the unemployed in the earlier period. Although the female unemployed were still relatively more numerous in the outer areas in 1984, the difference between the inner and outer conurbation in the male/female unemployment mix had narrowed very significantly. For the conurbation as a whole, the female proportion of the unemployed rose from 19.7 per cent in the late 1960s to 28.5 per cent in 1984.

Before leaving the question of the growth of aggregate unemployment it is important to underline the point that the growth is understated for three basic reasons. The first of these is the effect of people leaving the labour market altogether because of discouragement. The second factor is the change in the method of counting the unemployed introduced in 1983. Third, various special employment programmes have been introduced to mop up the growth in unemployment. In addition to these factors, of course, there is always some proportion of the unemployed who do not seek benefit support from the state and are, as a consequence, not counted as unemployed. Strathclyde Regional Council estimated in October 1984 that in addition to the 192 000 counted as unemployed there were 20 000 excluded owing to definitional changes, 26 000 unregistered unemployed, and 37 000 soaked up by the various special supplement measures (Strathclyde Regional Council 1984).

As the unemployment situation has deteriorated in the conurbation, the problem of long-term unemployment has become increasingly severe. Table 5.5 shows that the rise in long-term unemployment has been dramatic, particularly over the three years 1982–5. Over this period the numbers of unemployed males in Glasgow changed very little. Indeed the numbers fell slightly, although the basis of the count changed in between these two dates. However, the proportion of long-term unemployed rose from 41 per cent to 52 per cent (taking those unemployed for a year or more) or from 18 per cent to 36 per cent (considering those unemployed for two years or more). In January 1985 nearly 5000 of Glasgow's male unemployed had not worked for at least five years. The problem of long-term unemployment is generally less pronounced for the outer conurbation relative to the city of Glasgow, and for females relative to males. However, this latter phenomenon is in some measure a consequence of the less favourable social security provision for married women. The relatively low proportion of females unemployed in the long-duration category is a well-known national phenomenon (McGregor 1978).

The decade 1971–81 witnessed major changes in both national and local labour markets as well as a series of urban policy initiatives in many areas, including Glasgow. As we have shown, these changes included substantial employment declines (with a continuing fall in the relative significance of manufacturing employment) and a massive rise in unemployment, particularly in the latter part of the period. Within Glasgow, significant rebuilding and rehousing took place with increased public sector provision in the inner city and the rehabilitation of many older tenemental properties. Finally, a large number of area-based initiatives were mounted which included among their aims the reduction of unemployment (for example, Glasgow Eastern Area Renewal), or which were set up as a response to major job losses (for example, Clydebank Task Force and Enterprise Zone). It is within this context of labour market change and policy initiative that it is interesting to examine the stability or otherwise of the spatial structure of unemployment over time.

To provide a context for the detailed spatial analysis, the situation in the districts making up the conurbation is summarized for 1971 and 1981 in Table 5.6. Even at this fairly aggregated level, there is evidence of the great diversity of unemployment rates at each date. However, there appears to be a high degree of stability over time, at least in the ranking of the different parts of the conurbation. Between 1971 and 1981 Cumbernauld and Kilsyth dis-

Table 5.5. *Long-term Unemployment in the Clydeside Conurbation by Gender and Area, 1979, 1982, and 1985*

Gender	Year	Duration of unemployment		
		Up to 1 yr.	1–2 yrs.	2 yrs.+
Glasgow				
Males	1979	20 738 (67)	10 331 (33)	
	1982	32 366 (58)	13 036 (23)	10 160 (18)
	1985	26 553 (48)	9 540 (17)	19 037 (35)
Females	1979	9 358 (85)	1 712 (15)	
	1982	14 170 (72)	3 441 (18)	2 037 (10)
	1985	13 281 (65)	3 223 (16)	3 843 (19)
Outer conurbation				
Males	1979	16 475 (73)	6 039 (27)	
	1982	33 188 (68)	10 023 (20)	5 761 (12)
	1985	28 386 (54)	9 313 (18)	14 095 (28)
Females	1979	11 363 (82)	2 581 (18)	
	1982	18 794 (76)	4 217 (17)	1 796 (7)
	1985	15 406 (69)	3 379 (15)	3 676 (16)

Note: The figures in parentheses are raw percentages. In 1979 it was not possible to differentiate among those unemployed for over a year or more.

Source: MSC.

Table 5.6. *Male Unemployment Rates in the Clydeside Conurbation, 1971 and 1981*

Area	1971 rate (%)		1981 rate (%)		Proportional changes in rates	Absolute changes in rates
Eastwood	2.3	(317)	5.1	(767)	2.2	2.8
Bearsden and Milngavie	2.5	(241)	5.3	(581)	2.2	2.8
Strathkelvin	4.9	(1 018)	9.5	(2 337)	1.9	4.6
East Kilbride	5.1	(1 045)	11.8	(2 929)	2.3	6.7
Renfrew	7.5	(4 271)	14.3	(8 507)	1.9	6.8
Cumbernauld and Kilsyth	7.6	(982)	13.8	(2 478)	1.8	6.2
Hamilton	9.3	(2 748)	17.8	(5 468)	1.9	8.5
Motherwell	9.3	(4 120)	19.6	(8 124)	2.1	10.4
Clydebank	9.5	(1 600)	20.8	(3 023)	2.2	11.3
Monklands	10.5	(3 332)	22.0	(6 817)	2.1	11.4
Glasgow City	12.9	(34 897)	23.5	(49 176)	1.8	10.6
Glasgow City (outer)	10.7		23.0		2.1	12.3
Glasgow City (inner)	16.1		24.6		1.5	8.5

Note: The absolute figures are given in parentheses.

Source: Census of Population, 1971 and 1981.

trict exchanged places in the rankings with the district of Renfrew, but this was the only change of position. Consistent with this is the finding that the proportional change in unemployment rates showed very little variation by district over the decade. The proportionate growth in unemployment was lowest for the city of Glasgow. In the outer conurbation the labour force expanded against a decline in *in situ* employment. However, the key conclusion is the high degree of stability of the spatial structure of unemployment and the relative uniformity of the proportionate growth in male unemployment rates across districts. Within this there appears to have been a slight tendency towards a narrowing of unemployment differentials.

However, despite some narrowing of differentials, at both dates the city of Glasgow experienced the highest male unemployment rates. In 1971, by far and away the worst area was inner Glasgow with rates which were, at a minimum, 50 per cent greater than those prevailing in other parts of the conurbation. However, by 1981 the difference between inner Glasgow and the other high-unemployment areas of the conurbation (including the peripheral areas of Glasgow city) had narrowed substantially in both absolute and proportional terms. The peripheral areas of Glasgow city experienced an increase in unemployment which in proportionate terms was about the average for the districts making up the outer conurbation. It is true to say that the major element of change in the spatial configuration of unemployment over the decade was the relative improvement in the position of the inner parts of the city of Glasgow, although this area still had the highest rate of male unemployment in the conurbation in 1981.

In the rest of this section we concentrate on the differential experience of the inner and outer parts of Glasgow over the decade. In 1971 there were heavy concentrations of the worst areas in the inner city, particularly in the East End. By 1981, these inner city concentrations of areas of high unemployment had been broken up, if not completely eradicated. However, new concentrations had developed in the city's peripheral public sector housing estates, particularly in the north-west (Drumchapel) and the north-east (Easterhouse–Garthamlock).

A number of points can be made at this stage. First, the change observed may be indicative of the effectiveness of the Glasgow Eastern Area Renewal (GEAR) project, although it cannot throw light on *cost* effectiveness or, indeed, on spillovers and other distributional consequences. Second, new concentrations of high-unemployment areas have emerged on the periphery. There is no iron logic to this process. The breaking up of concentrations of male unemployment in the inner city could have been accompanied by a dispersal of the worst areas throughout the city rather than by the emergence of new concentrations.

Third, the process of change is clearly linked with major developments in the housing sector taking place in the city of Glasgow between 1971 and 1981. The inner city concentrations of areas of high male unemployment were predominantly to be found in neighbourhoods with a high proportion of rented tenemental properties. In 1971, only ten of the worst thirty areas were dominated by public sector housing. Many of these tenemental areas were cleared or improved with consequent changes to the level and structure of their indigenous populations. By 1981, of the worst thirty areas for male unemployment, twenty-five were in areas dominated by public sector housing.

In seeking to explain why unemployment rates vary between different parts of a single city or conurbation, two broad lines of argument are generally employed. The first argument is that neighbourhoods differ in their population characteristics and this gives rise to unemployment variations. Thus, those parts of the city with, for example, high proportions of unskilled workers will have relatively high unemployment rates because the unskilled *in general* are more at risk of unemployment than their skilled counterparts. The second argument is that the residents of high-unemployment areas face structural barriers to successful labour market participation. These would include distancing from employment nodes, problems of labour market information (McGregor 1983), and discrimination against certain neighbourhoods. Of course, these lines of argument are not necessarily exclusive.

The change in the balance of unemployment between the inner and outer areas of the city of Glasgow can be considered within the framework provided by these two lines of argument. The narrowing of the differences could be consistent with one or more of the following:

(*a*) A reduction in the differences in the labour-force characteristics of the populations of the inner and outer city
(*b*) An increase in the proportion of the city's jobs which are within reasonable travelling distance for the residents of the inner city
(*c*) A decline in the size of the inner city's labour force relative to that of the outer city
(*d*) An improvement in transportation systems leading to a reduction in spatial segmentation of the city's labour market
(*e*) An improvement in labour market information systems leading to reductions in informational inequalities across the city.

Policy Evaluation

While there have been massive changes in the economy of the Clydeside conurbation, policy to deal with these changes, or their adverse consequences, can be characterized in two ways: firstly, there are policies which were developed in the economic context of the 1960s and have been extended into the 1970s and 1980s; secondly, with the increasing realization that the nature and extent of the economic malaise was much more serious, new policies and approaches have been devised, not to replace the older policies, but to augment them.

The policies which were developed, or enhanced, in the 1960s were based on an assumption that the national economy was reasonably healthy and that the problem to be addressed was the spatial distribution of investment, output, and employment which generated unacceptable differentials between locations in the rates of unemployment and participation, and in levels of output and income. The major component of such policies was the spatial redistribution of employment opportunities away from the congested South and Midlands and towards the North and West. It was possible to justify such policies, and the costs which they might impose upon companies by forcing them into the relocation or multi-plant operations, both in terms of 'equity' (that is, the natural justice of trying to ensure that all people had roughly equal chances of acquiring work) and in terms of 'efficiency' (that is, maximizing output at the national aggregate level). Such enhanced levels of output would be achieved by easing 'tight' labour markets in the South and Midlands where labour shortages were pushing up local wage rates, by reducing the pressure of demand for space in those regions where rents were rising rapidly, and by the avoidance of other negative agglomeration economies such as traffic congestion. Such policies were predicated upon an assumption that there was growth in the national economy which could be relocated provided that the governmental cost of doing so (in subsidies etc.) was more than recouped in lower unemployment and related welfare payments (Cameron 1979; Marquand 1980).

By the late 1970s, these assumptions clearly no longer held. Whereas registered unemployment and total vacancies (as opposed to registered vacancies where there is serious under-recording) had remained of roughly the same order of magnitude until 1978, unemployment climbed rapidly after that date. Thus, the growth of demand-deficient unemployment meant that it was no longer possible for policies solely to concern themselves with the matching up of labour surpluses in the North with job surpluses in the South, with or without retraining (Lever 1986). The recession had a number of causes, including the overvaluation of sterling, a worsening of the comparative cost performance of the UK economy relative to industrial competitors, and the restrictions on public expenditure at national and local levels. Its translation into a national unemployment rate was exacerbated by a growth in the population of employable age as a number of school-leavers exceeded the numbers reaching retirement age, and the tendency for expanding sectors to employ workers who had not hitherto been registered as unemployed (Hughes 1981; Townsend 1983).

New policies had to be developed to contend with the new emergent problems such as the long-term unemployed, the very high rates of school-leaver unemployment, and the creation of jobs by enterprises which were unlikely to be economically viable in the long run (Fothergill and Gudgin 1982, 170–87). Nevertheless, the old policies were continued, albeit with some modifications, although they were asked to operate in a different economic context. Regional policy, for example, relied much more heavily upon incentives, while the control of industrial development in the South and Midlands through the use of Industrial Development Certificates largely ceased.

Older Policies

Regional aid

An analysis of grants awarded under the Regional Development Grant (RDG) system in the four-year period 1979–83 shows several broad trends. During that period, RDGs worth approximately £100 million were made to firms in the Clydeside conurbation. After a 60 per cent rise in 1980–1, the grants declined in real terms as the recession deepened. Analysis of the grants indicates that much of the money was used in maintaining existing employment in many of the older sectors such as steel and heavy engineering. Expenditure was concentrated in large-scale capital-intensive investment and replacement investment with only small employment gains. Over the conurbation as a whole, grants per worker in 1979–83 averaged £310; the only sectors with assistance levels significantly higher than this were food, drink, and tobacco manufacturing (£640), metal manufacture (steel) (£640), mechanical engineering (£520), and instrument engineering (£550). In absolute terms, the largest grants were made to food, drink, and tobacco (£22 million), mechanical heavy engineering (£26 million), and steel (£15 million). The majority of the grant, therefore, was used to modernize the old industrial base of the region. Potentially new and expanding sectors within chemicals, electrical engineering, and vehicles (including aerospace) have not benefited significantly from RDG, although the small instrument engineering sector did do well on a per-worker basis despite the small absolute amount of grant (£2.9 million).

Although unemployment rates in Glasgow have been persistently higher than elsewhere in the conurbation, the majority of the RDG expenditure has gone to the outer areas. Overall, 68.7 per cent of RDG expenditure was incurred in the outer city which has 45 per cent of the manufacturing employment, whereas Glasgow, with 55 per cent of the employment, received only 31.3 per cent of the assistance. This ratio appears to have remained fairly constant over the four-year period. It is possible that this imbalance merely reflects the different economic structures of Glasgow and the outer conurbation. Grants tend to be claimed when expansion or investment is contemplated, and if Glasgow has a disproportionately low share of expanding industries then its share of RDG will be lower. However, in almost all sectors the grant per worker was higher in the outer city than in Glasgow. Overall, the grant per worker was £420 in the outer city and £200 in the inner city in the period 1979–83. Only in three sectors—instrument engineering, vehicles, and the residual miscellaneous manufacturing—did the grant per worker in Glasgow significantly exceed that for the outer city. The other sectors where Glasgow's grant was marginally higher are the trivial cases of miscellaneous metal goods and textiles. The recent White Paper on regional development (Department of Industry 1983) suggested that the mean cost per job created or maintained by the RDG is approximately £35 000. In this case, then, we might expect some 3000 jobs to have been created or maintained in the Clydeside economy between 1979 and 1983.

Regional Selective Assistance (RSA) brought £36 million to the Clydeside conurbation between 1979 and 1983, of which the largest sectoral recipients were heavy mechanical engineering, electrical engineering, and food, drink, and tobacco, but not, unlike with RDG, the steel sector. The imbalance between the inner and outer cities was more acute for RSA than for RDG. The inner city received only 22 per cent, despite having 55 per cent of the conurbation's total employment. Of the £36 million allocated, £21 million was assigned to job-creating projects and £15 million was assigned to job-maintaining projects. Because job creation and maintenance figures must be provided by applicants under the scheme, it is possible to estimate that from RSA expenditures some 16 600 jobs were created or preserved in the conurbation between 1979 and 1983.

The third element in conventional regional policy is the scheme of assistance under the Office and Service Industries Scheme (OSIS). In 1979–83, compared with £137.1 million to manufacturing, it is clear that the service sector, which despite its capacity for employment growth received only about £7.0 million attracted relatively little assistance. Under the provisions of the OSIS, expenditure on assistance must not exceed £8000 per job created, thereby giving a figure of approximately 900 jobs; however, in this case the balance of advantage spatially does lie with the inner city, as much of this assistance went to establishments in and on the periphery of Glasgow's central business district.

The attraction of inward investment

A major part of regional policy in the 1960s and early 1970s was the use of financial incentives to attract employment from the South and Midlands, and from abroad, to depressed regions such as Clydeside. This would occur either in the opening of branch plants by companies managed from outside Scotland or, much less frequently, in the total movement of firms to the depressed regions. The ability of local authorities in Clydeside to take part in industrial development, including the attraction of mobile investment, stems from a number of Acts including the Town and Country Planning (Scotland) Act of 1947, the Local Government (Development and Finance) (Scotland) Act of 1964, and the 1946 New Towns (Scotland) Act. The main incentives offered by local authorities (in addition to some regional policy aid) included

Table 5.7. *Incoming Employment in Glasgow and West Central Scotland, 1945–1970*

Period	Glasgow		Rest of West Central Scotland	
	Maximum	Difference between actual and expected	Maximum	Difference between actual and expected
1945–51	8 330	−15 970	20 040	+16 700
1952–9	7 040	−2 200	8 150	+7 840
1960–5	1 100	−10 910	13 160	+8 040
1966–70	1 880	−5 020	7 480	−170
Total 1945–70	18 350	−34 100	48 830	+32 410

Source: Henderson (1980).

the provision of serviced industrial sites and advance factories, the offer of 75 per cent grants towards the cost of constructing factories, the provision of basic infrastructure such as roads, environmental improvement packages, and 'key-worker' housing (Henderson 1980, 182–3).

The numbers of jobs brought to Glasgow and the rest of West Central Scotland are shown in Table 5.7, expressed as the maximum employment achieved at any point in time up to 1970. By 1970, decline had set in so that actual totals in 1970 were about half the maximum value. It is clear that Glasgow attracted immigrant firm employment at a much lower rate than the outer conurbation throughout the period 1945–70 although the disparity was least in 1952–9. The comparison between actual values and 'expected' values is based on an expected value calculated by allocating all immigrant employment in Scotland to areas in the ratio of their manufacturing employment. On this basis, Glasgow consistently does badly, whereas the rest of the conurbation does well, at least up to 1965. Henderson argues that this disparity is attributable to: firstly, the creation of the two New Towns, East Kilbride and Cumbernauld, which account for about 13 000 of the 32 000 additional jobs in the outer conurbation; secondly, the poor image of Glasgow in environmental and labour terms which deflected inward investment; and, thirdly, the fact that the cost advantages enjoyed by Glasgow in terms of transport, linkages, and labour were largely dissipated after the mid-1950s.

By the mid-1970s, two clear trends were apparent. Firstly, the national and international recession which developed after 1973 meant that there was much less mobile industry within Britain, and international investment was increasingly likely to go to low-cost labour areas such as the newly industrializing countries. Secondly, the fact that all local authorities had powers to undertake promotion and attempt to attract inward investment was proving confusing to potential investors—especially those abroad—and in consequence might well have been proving counter-productive. At least as far as Scotland was concerned, the answer after 1978 was simple—to co-ordinate all promotional activities through a single channel, the SDA. By 1981, this had been formalized into Locate in Scotland (LIS) as a joint venture incorporating the promotional activities of the SDA and the financial incentives operated by the Industry Department for Scotland who equally provide the resources (Young 1984). LIS's main role is overseas promotional work, with a budget of about £2 million per year operating through offices in New York, San Francisco, Chicago, and Houston, and with an agency in Tokyo. LIS now claims that about 9000 prospective new jobs in new companies and 10 600 additional jobs in existing companies have been brought to the whole of Scotland. LIS does appear to provide a quick and effective service, although it has been criticized for being too reactive, lacking defined targets in terms of firms or sectors. Sectorally, analysis of the inward investment flows shows a marked concentration upon electronics and mechanical engineering and spatially the United States of America has provided the majority of immigrant firms.

The major problem with LIS, from the perspective of the local authorities which it in effect replaced, is its spatial effect within Scotland. LIS, in general, markets Scotland as a whole rather than particular locations within Scotland, although it is currently considering special promotions for the Scottish Enterprise Zones because of their special tax incentives. In practice, LIS responds to the preferences of its customers and thus recognizes the attractions of certain locations, such as the New Towns, showing such places to business visitors who express an interest in Scotland. LIS feels that the inward investor is, for practical purposes, only likely to be attracted to an area-initiative location at a late stage in that initiative when most of the environmental problems have already been solved. Indeed, many of the companies coming to Scotland have themselves moved out of inner city areas in the older parts of the USA. Such considerations therefore raise question marks over the likely success of LIS in attracting investment into areas such as the GEAR scheme in Glasgow's East End or the initiative areas such as Motherwell or Coatbridge. Thus the problems which led to inward investment avoiding Glasgow and some of the older industrial towns when local authorities were responsible for promotion and industrial attraction may still be of concern within the overall effect of LIS (Moore and Booth 1984).

Small firms

Policies supporting small-firm development have grown steadily in importance over the last decade as the nation's poor economic performance has been partly attributed to the low percentage of total output from small firms and to the continued decline of employment in larger enterprises (Hood

1984). The Clydeside conurbation does have a large stock of small businesses, although a long period of urban redevelopment has had a serious effect on this economic base (Bull 1981; McKean 1975). Small-firm promotion, in conjunction with environmental improvement and the provision of small industrial premises in the urban cores, has been the main element in the inner city revitalization programme (Hood *et al.* 1982). Small-firm development in Clydeside has, therefore, an unusually wide range of support agencies with differing forms of advisory and financial support from national programmes of the Department of Industry, through the SDA, through local authorities, and increasingly through specialist agencies linking the private and public sectors. Small-firm support is also becoming increasingly spatially focused in GEAR, the Joint Economic Initiatives, and other area-based schemes. A broad range of assistance is covered, comprising financial support, advisory services, management and labour training, and premises. This diversity raises the potential problem of overlap and confusion, although this danger does appear to have been minimized by co-operation between agencies which have developed interlocking rather than overlapping functions. For example, the Small Business Division of the SDA provides only specialist technical and investment advice, leaving the provision of general advice on setting up in business to Glasgow Opportunity, the local Enterprise Trust. Similarly, while much of the finance for small firms is provided by the Small Business Division of the SDA, this is complemented by the local authorities' powers to offer loan finance to enterprises which are considered too marginal or too risky for the Small Business Division to contemplate.

The proliferation of agencies has created a level and a diversity of services to small business which no one institution would have been able to finance or to provide. Increasingly, too, the private sector has become involved in offering its expertise, and by seconding staff through agencies such as Glasgow Opportunity. The diversity of agencies has achieved a high level of penetration, with approximately 70 per cent of small businesses aware of the SDA's programme of assistance and around 30 per cent aware of local authority schemes. At present, the system of advice to small businesses appears capable of development in three ways. Firstly, increasing staffing would enhance the level of services currently offered by several agencies. Secondly, there is a feeling that assistance tends to be directed either at new firms start-ups or at small businesses which get into difficulties, whereas more effort, given additional resources, might be directed at established, secure, small businesses which have the capacity, were they given more help, to expand output and especially employment. Thirdly, less formal businesses—community businesses, co-operatives, and other forms of worker-managed enterprises—offer new and expanding outlets for the providers of small-business assistance.

New Policies

Alternative enterprise

If conventional policies provide only a partial response in managing economic transition because they lack a finely targeted focus on the most disadvantaged, is there an alternative 'bottom-up' approach which can help bridge the gap between the market and the dispossessed? In Clydeside, a new wave of enterprise has arisen alongside the more traditional alternatives such as workers' co-operatives. These new forms of enterprise are based on collective ownership and control, but, rather than restrict this to workers as in co-operatives, they attempt to involve the wider local community. These enterprises have been called community businesses. Here we want to observe how this option for enterprise at the community level has emerged and how it might be developed.

Community businesses have been defined as:

> a trading organisation which is owned and controlled by the local community and which aims to create ultimately self-supporting and viable jobs for local people in its area of benefit, and to use any profits made from its business activities either to create more employment or to provide local services or to support local charitable work. A community business is likely to be a multi-purpose enterprise and may be based on a geographical community or on a community of interest. It will have limited liability and in some cases will acquire charitable status. (Community Business Scotland 1981.)

This highlights a number of essential principles. Firstly, there is the principle of *community ownership and control*, which should be institutionalized in the very structure of the enterprise by opening up membership to as wide a section of the local community as possible, and ensuring maximum participation and accountability in the management of the enterprise. Secondly, there is the principle of *targeted employment creation*, which seeks to meet the economic and social needs of a particular locality and its residents. In employment terms this means trying to create job opportunities specifically for local unemployed people. Thirdly, there is the principle of *collective profit*, whereby any surplus wealth generated by community enterprise is ploughed back into the community, either in the form of communal services or in the creation of more employment opportunities. Fourthly, there is the underlying principle of *synthesis*, bringing together goals of community development and economic regeneration. Community businesses are attempting to break down old modes of compartmentalized thinking which separate business activities from community needs.

While these four basic principles can be considered to embody an embryonic theory of community business, attempts to put them into practice have had variable success. Firstly, from the democratic perspective, community businesses have not progressed as far as some of the proponents had hoped in securing community commitment. Indeed, some commentators have suggested that the goals of local control and targeted employment creation might be considered 'discretionary' objectives (Hayton 1983). Most Clydeside community businesses have a low level of membership. The presence of apathy or cynicism should not be surprising given the unfavourable environment for enterprise, for example, on the peripheral estates.

The process of starting up generally involves piecing together disparate resources from a variety of agencies and itself can lead to disintegration and loss of enthusiasm. Once trading has begun, the emphasis in many community businesses centres on gaining commercial credibility, and the objectives of democracy and accountability take second place to market goals. Dependence on public agencies for support tends to reinforce the search for business credibility.

Secondly, the economic impacts of community enterprises have been marginal to the scale of the problems facing the communities themselves, although some positive contributions have been made in providing employment and generating new economic activity. The number of jobs associated with these businesses is small. In a study of twenty-five operating enterprises on Clydeside it was calculated that between 400 and 800 jobs had been generated, including part-time employment and jobs dependent on temporary public schemes such as the Community Programme (McArthur 1984). For those individuals the benefits may be highly significant, but in terms of the scale of the unemployment problem the overall impact can only be described as marginally beneficial. However, any fair assessment would need to account for multiplier impacts such as enhanced local spending power, whether this is in the form of self-supporting jobs or the leverage of new public investment into these communities via the community enterprise.

Qualitatively, the bulk of full-time employment tends to be concentrated on fairly low-level semi-skilled jobs reflecting reliance on Manpower Services Commission (MSC) support. While there are good examples of specialized craft-based activities, the principle of targeted employment undoubtedly limits the type of commercial ventures open to these businesses.

Finally, the distributional impact of employment generated by community businesses needs to be more fully considered. In particular, the benefits for particular segments of the local labour market need to be analysed. While community businesses might claim success in reaching certain sections of the disadvantaged such as the longer-term unemployed, they are less successful in providing full-time employment opportunities for women (McArthur 1984).

If the impact of community businesses in regenerating local economies and creating employment is limited, is there a case for giving them public support? Firstly, more conventional models of economic development have consistently failed to meet the needs of the most disadvantaged. Created by massive public planning and left untouched by the market, the peripheral housing estates of Glasgow are beginning to take the first steps towards combating years of neglect. The absence of a developed economic infrastructure makes them unsuitable for major market-oriented programmes and strategies directed at 'areas of potential'.

Another policy option would involve major training and retraining programmes, focusing on the individual within the labour market rather than the community. While targeted training is an important dimension of policy for enterprise and enhances individual mobility, it does not in itself create new job opportunities. If demand for labour is depressed, training serves to redistribute unemployment. At the same time, this approach does not address community needs for jobs and services. Thus community businesses do have a role to play in the transition of the local economy because they reach out to areas and people untouched by sectoral strategies and small-firms development.

This is not a recommendation for blanket support by public agencies. A number of conditions need to be satisfied before a community enterprise can be considered viable even within its own terms. Firstly, there needs to be a demonstrable commitment within the community to support the objective of starting up a community business. In the Highlands and Islands the regional development agency supports community co-operatives by matching locally raised starting capital with grants and subsidized loans. This may not always prove viable in disadvantaged urban areas, although this argument needs more rigorous testing given the ability of remote rural communities to raise the necessary finance. However, some surrogate for financial commitment should be possible—for example, membership subscriptions.

The community enterprise structure also needs to be based on the principle of maximum participation and accountability. Strategic decisions should be taken by the membership and not usurped by individual activists or managers. If in this context it is decided in the short term to emphasize commercial objectives over democratic goals, this would be a decision of the whole membership. Otherwise a community business is no different from a conventional private sector enterprise where shareholders play only a formal role in decision-making.

We suggest that community businesses need to address two key distributional issues. The first is the degree to which employment creation should be targeted on locally unemployed people. Importing skills can enhance the efficiency and commercial success of an enterprise and, in the long term, create more local jobs. At the same time, however, types of business activity need to cater for the needs and capacities of local people as well as potential markets.

Secondly, if the benefits of both community participation and employment generation are to be truly extended, then issues such as the role of women need to be much more seriously addressed. There is a clear tendency to perpetuate existing gender stereotyping and keep women in conventionally accepted areas of activity such as 'home working' which offer low-paid and part-time employment opportunities. The failure even to acknowledge this as an issue of immediate concern means that no positive steps can be taken to ensure that women fully participate in the formation and development of broadly based community businesses. At the same time, opening up opportunities for better-paid employment for women would have positive economic multiplier effects in the community.

The responsiveness of the external policy environment to the needs of community business is also a critical issue. In Clydeside, community businesses have received public support from a number of agencies, but this has often been fragmented and inappropriate. There are conflicts between the market-based approaches of agencies such as the SDA and banks and the social objectives of local authorities. Available programmes of support operate to rules and timetables which fail to recognize the objectives or needs of community businesses. For example, Urban Aid applications go through a two-stage process of approval with both local authority and central government sanction required. Secondly, the MSC's Community Programme makes it difficult for projects looking to build on temporary schemes to provide permanent employment in wealth-creating activities to win approval. The decision to create a new regional support agency in Strathclyde to co-ordinate and target support has been backed by several public and private sector bodies. The problem of striking a balance between commercial and social objectives will remain, but this is the essence of the challenge of community business.

Self-employment

Given the depressed state of the employment market, it is natural for attention to turn to the possibilities of self-employment providing a short-run palliative for the unemployment situation, and possibly a long-run solution. National and local agencies have placed stress on the role of small businesses in providing an engine for economic development. Recently, the MSC has become strongly involved in this area through a series of self-employment training programmes and with the introduction of its Enterprise Allowance Schemes (EAS) on a national basis in August 1983.

The scheme is essentially a wage subsidy, providing a previously unemployed person with £40 per week for a year to set up his or her own business. By providing a guaranteed basic income above which the entrepreneur is able to keep anything else earned, it overcomes the perennial problem that those in receipt of unemployment or supplementary benefit face when they become engaged in income-generating activity. In most of these cases the unemployed can keep only a small proportion of any income they can earn without incurring benefit loss.

The difficulties attached to expanding self-employment among the unemployed are evident in the implementation of the EAS in Glasgow. In general terms, the scheme's uptake has been slow compared to other areas, as one might expect given the tradition of large-scale enterprise in the city. However, when one considers the intra-urban pattern, significant differentials emerge. The uptake of the EAS has been particularly low in areas where the unemployment problem is most severe. The MSC has reported a low uptake of the scheme in the peripheral council housing estates—precisely the localities where the concentrations of high and long-term unemployment are greatest.

In addition to high unemployment, a number of distinctive local factors inhibit the expansion of self-employment in the peripheral estates. Low levels of home-ownership reduce the possibility of local residents raising capital through home mortgage to launch a new business. Market opportunities for small construction firms are reduced because the bulk of household repair and maintenance work is carried out by local authority workers. This poses a particular problem as 'construction' accounts for a very large proportion of new firms started by unemployed people. Low incomes and the large number of householders depending on social security maintenance depress demand in the local economies and serve to reduce opportunities for new enterprise to start and flourish. Compared to other parts of the city, there is also an acute shortage of small cheap premises to house business starts. Furthermore, the large-scale investment in serviced sites and premises which has been the mainstay of urban policy since the mid-1970s has focused heavily on the inner conurbation.

When one considers the personal characteristics of EAS participants, the scheme appears to be relatively successful in reaching a significant proportion of the long-term unemployed. The MSC has conducted its own internal evaluation of the EAS. An assessment of participants six months into the scheme found that 25 per cent had previously been unemployed for more than twelve months. While clearly the scheme was having some impact on a major problem group nationally, even on the optimistic assumption that a similar ratio of long-term unemployed were participating in Glasgow, the impact on the disadvantaged will again be selective. In the city of Glasgow, 50 per cent of the male unemployed had been out of work for over twelve months in 1985. But even more worrying was the 35 per cent who had been jobless for two years or more: this hard-core group of very long-term unemployed will have even less chance of being reached.

One of the main inadequacies of the EAS as it stands is the lack of a formal training component despite the obvious need for some input in this area given the client group in mind. This largely reflects the scheme's genesis within the Employment Division of the MSC as opposed to the Training Division. However, in Glasgow the MSC has recognized the lack of training as a problem likely to affect the prospects of business success and have funded a pilot training course for EAS participants. Participation is voluntary and courses are run during evenings over a six-week period. Organizers report that the types of people taking the courses often have very poor educational qualifications, vague business ideas, and few skills—precisely the type of people in need of long-term hand-holding if they are ever to have a chance of running a business efficiently. Given the cultural constraints on entrepreneurial capacity emerging out of Glasgow's tradition of large-scale heavy engineering industries, training programmes are a necessary prerequisite to any expansion of self-employment among the unemployed. The MSC's provision here is expanding. A 'training for enterprise' category has been introduced as part of the Job Related Programme under the new Adult Training Strategy. It is likely that most training initiatives in the self-employment area will in future depend in whole or part on MSC resources.

The MSC has funded a number of low-level self-employment courses in the Glasgow area, some run through local technical colleges. They have tended to be of relatively short duration, lasting between eight and twelve weeks, with participants paid an allowance. Course curriculum is mixed. The emphasis can vary between enhancing the 'production' skills of potential entrepreneurs or providing them with the 'business' skills necessary to set up and manage a new enterprise. Courses are geared to launching a business start at the end of the training period, and participants often begin preparing for this during the course.

Organizational Capacity

Just as we distinguished between old policies which are now confronting a different economic climate and newly devised policies initiated to augment the older ones, it is also possible to distinguish between old agencies of policy delivery and relatively newly created ones. The most obvious contrast here is between the structure of local government authorities which have acquired increasing powers of industrial development and the SDA, created *de novo* in 1975.

Local authorities

There are three distinctive elements in the strategic decision-making process of local authorities in their economic development programmes. At the basic level, there is the decision to engage in a programme of 'hardware' or physical development. This is concerned with planning for the provision of industrial land, building factories and other premises, and making loans or grants for environmental improvement. Practically every authority has some involvement at this level.

Secondly, there is the decision to become involved in the 'software' of business development. This is less common, and involvement varies significantly. While most authorities offer some form of business advice, other forms of support such as training, business planning, and finance present greater problems in terms of political commitment and capacity.

Thirdly, the most sophisticated level of activity involves the decision to produce an economic development plan within the corporate policy framework of the authority. Adopting a rational planning approach to economic policy is largely undeveloped. Only one local authority within the conurbation has actually produced a statement, although other authorities have moved towards strategic programme documents for key elements of their activities.

Hardware: Factory development Within the format of programmes directed towards local economic development, the provision of industrial land and factories is the major item of capital spending and is therefore a critical element in the development strategy. Larger authorities such as Glasgow District Council (GDC) are spending annually about £4 million on currently programmed developments. In the context of total authority spending this is not especially significant, but in the context of expenditure on economic development the provision of sites and premises is the central resource user.

The overwhelming strategy of district authorities in terms of factory provision is to meet perceived gaps left by other providers at the local level. This has translated into a programme of small-unit and workshop provision, sometimes involving more 'sheltered' accommodation. While the planning of factory provision is not based on highly commercial appraisals of rates of return and profit maximization, authorities are looking generally to cover costs over the medium term. They do not want to build units that will remain unoccupied. They concentrate on the smaller developments which are normally easier to fill and where the investment risk is evenly spread.

Project appraisal is also based on some notion of social benefits likely to accrue from building units in certain areas. The choice of site is ultimately a political decision. This is most clearly seen in the decision of GDC to build small units in the peripheral housing estates to support its own priority area development strategy. It is acknowledged that in many cases such provision will not be a strictly commercial proposition, but the object is to stimulate demand through provision allied to other business development programmes aimed at creating an interest in self-employment or co-operative forms of enterprise. In some cases it is hoped that provision will attract businesses from outside the locality, but the primary object appears to be the creation of indigenous enterprise within the context of a wider community regeneration policy.

The factory programme of East Kilbride Development Corporation (EKDC) stands in marked contrast to this general strategy of small-unit provision, reflecting a basic difference of approach in terms of business development policy. With the districts, the focus on small units reflects a broader economic programme based on the development of small local enterprise. Business advisory services, financial assistance, and other programmes are geared towards the perceived needs of the small firm. In contrast, while EKDC does not underestimate the importance of providing for local enterprise, its development strategy is still based on winning major inward investments. Even with an acknowledged limited amount of mobile investment, it is felt that one major incoming firm or expansion can be more significant in employment terms than a large number of small firms.

District authorities look to small firms because they acknowledge that they are unlikely to attract major inward investment. This is partly because of the attractions of New Towns such as East Kilbride and partly because they feel that LIS steers firms to New Towns rather than older urban areas.

Software: Business advice and finance All authorities provide business advice, although sometimes this is limited to help with planning applications or passing on enquiries to other specialist agencies. Officials in the smaller districts designated as Business Development Staff are invariably located in the planning departments and are usually trained planners. Larger authorities employ a number of staff in special units which may or may not be part of their planning departments. In Strathclyde Regional Council, for example, the Industrial Development Unit (IDU) is located in the Chief Executive's Department.

Some authorities, such as Glasgow and Strathclyde, attract people from the private sector to their economic development staff, but this is the exception, not the rule, among local authorities. This general lack of private sector experience could be seen as a gap which it would be fairly simple to fill. Improving the training of existing staff, appointing external specialists, or using other agencies, such as the Lanarkshire Industrial Field Executive or Enterprise Trusts, are some of the ways this problem could be overcome.

Smaller councils, in short, offer essentially a signposting service. Even larger authorities, like the regional councils, see their role as limited. Strathclyde's IDU, for example, has not been geared up to playing an extensive supporting role for small enterprise and the council therefore supports the work of other agencies such as the SDA, Enterprise Trusts, and the Scottish Co-operative Development Committee which can provide additional and more specific help.

In terms of providing finance for enterprises, most local authorities do not see themselves as having an important role to play beyond fairly small grants and loans. It is interesting to compare this attitude with that adopted by some English authorities such as the Greater London Council, West Midlands County Council, and Sheffield. These councils have taken a more systematic interventionist role which includes direct investment in businesses. There is a critical difference in the context of English and Scottish local government in that there is no equivalent of the SDA in England. Clydeside councils see it as the SDA's responsibility to support the growth of new business through direct investment.

As a result, most councils have no explicit guide-lines governing their loans and grants, but react in an *ad hoc* manner. The only exception to this general rule is Glasgow, which has adopted a set of written guide-lines specifying the type of firms it is prepared to support, together with the conditions and forms of aid it may provide. These conditions include a clause relating to an acceptable employment policy and the preferred recruitment of workers from disadvantaged areas. The council makes it plain, however, that these conditions are flexibly applied. They act more as a political statement of intent than a tightly defined policy.

Adopting rational planning models Only GDC among the district councils adopts a clearly laid-out economic development plan, embracing both a brief analysis of developments in the local economy and a range of programmes designed to help in the regenerative process. Other districts have a series of programmes, but these are not explicitly linked together in a comprehensive statement of objectives and policy initiatives.

Smaller district authorities argue that it is not viable for them to produce detailed economic development plans and that the Local Plan Statements provide an adequate vehicle for setting down proposals for development. For small district councils a fundamental question is: What is the point of producing economic development plans at all? They recognize that their role is marginal in many respects. The local economy does not function simply within district boundaries, but is significantly influenced by conurbation-wide phenomena. The implication of this is that only the city- or region-wide authorities such as Glasgow or Strathclyde have a sufficient territorial responsibility to begin to be able to make a significant impact through positive policy measures. As soon as this city-wide level is reached, however, disputes soon emerge about how far economic development strategies should be related to other aspects of corporate planning, and in particular to strategies designed to counter deprivation.

Local authority economic development policies reflect the needs of existing industrial areas and the demands of industrialists. They also seek to complement the efforts of other agencies. Programmes are based on realizing the potential of areas considered to be most attractive to industry, and this largely excludes areas of deprivation. An argument in favour of concentrating resources on established industrial areas is that these are major localities of employment serving a much wider area. However, if the residents of deprived areas are to benefit from the employment provided, then issues of training and mobility need to be more systematically addressed and programmes more targeted in favour of such people.

In contrast to the generally reactive approach of most district councils, the region has an integrated strategy for combating deprivation, embracing both economic and social policies. It has identified a number of localities as Areas of Priority Treatment (APTs) on the basis of established indicators of social deprivation. Following this spatially based approach to community regeneration, it has also identified a number of areas for the purpose of Joint Economic Initiatives (JEIs) which provide for a range of targeted programmes mounted in conjunction with the SDA and district councils. These JEIs do not necessarily coincide with APTs. Indeed, the criteria for selecting an area for a JEI are significantly different from the criteria employed for defining APTs. This partly reflects the fact that JEIs are *joint* initiatives, and, in particular, involve the SDA making a significant input of resources.

The method of JEI prioritization may be questioned. First, the indicators of deprivation used were based on employment exchange areas which, because of their widely drawn boundaries, managed to exclude some of the worst pockets of unemployment. This led the council to add a further five areas to the list of twenty-three areas provisionally identified.

A number of 'opportunity factors' were then applied to this list. These included whether the areas had a basic infrastructure suitable for economic development, such as land and the presence of existing small businesses. As a result of this exercise, a number or areas were excluded from further consideration. This left a short list of twelve areas. In this process one factor considered was whether areas were adjacent to existing special initiatives. An assumption was made that such areas would be excluded because they would benefit from a 'trickle-over' effect and would not therefore need special help. This has yet to be proved, and a number of factors could preclude its occurrence. The degree of spillover would depend upon the skill levels, mobility, and social support (for example, under-fives provision) in the area adjacent to the initiative.

Perhaps the most penetrating criticism relates to the inclusion of opportunity factors in the first place. What they effectively measure, in many respects, is the product of previous local authority decisions. The pattern and quality of an area's economic base is highly dependent upon public provision. Educational services, transport services, and, most importantly, housing will have critically affected an area's labour-force characteristics and communications infrastructure. Similarly, the quality and availability of industrial sites for development or occupation will have been determined by the past actions of the local authorities and the SDA.

The insistence of public authorities on the inclusion of opportunity factors is a product of the severe financial and political constraints upon them. It is indicative of the general problems associated with discriminatory investment. However, more than this, it is also highly dysfunctional in terms of the economic regeneration of the most disadvantaged areas. When local authorities can only entertain the possibility of joint action and positive discrimination in traditional industrial and commercial zones, it renders large areas of the city unprovided for, since they have no infrastructure suitable for development. The logic of this approach means that existing inequalities are reinforced.

Despite this emphasis on existing industrial areas, the need to direct some resources to the economic development of deprived communities is acknowledged. GDC's priority areas strategy is looking at these localities—in particular, the great peripheral housing estates—after having concentrated efforts on inner city areas such as GEAR and Maryhill. There is clearly a long way to go if GDC is to achieve the objective of creating identifiable communities with the range of social and economic facilities and opportunities of a medium-sized town (GDC 1978), and the critical thing as far as the authority is concerned must be securing the commitment of other agencies, in particular Strathclyde Regional Council (SRC) and the SDA. Some moves have been made to this end, but they are tentative and do not relate well to the scale of the problems needing to be tackled (GDC–SRC 1985).

The SDA

Within the UK, central government has developed a variety of packages designed to stimulate regional economic development, primarily in order to reduce spatial imbalances in investment and unemployment. Special-purpose agencies have also been created—for example, the Industrial Reorganization Corporation (1966–71) and the National Enterprise Board (NEB)—to develop sector- and industry-based strategies cutting across regions. The first specifically regional development agency in the UK was the Highlands and Islands Development Board created in 1965. This provided a model for the SDA when it was established under the Scottish Development Agency Act of 1975.

The SDA has responsibility for developing the Scottish economy through financial assistance to businesses and provision of industrial infrastructure. It combines elements of the old NEB in its investment and sectoral strategies with the activities of an industrial estates corporation. Additionally, the agency has resources for improving the urban environment in association with local authorities.

Since 1975 there have been no major functional changes to the agency, although the emphasis of its activities has shifted. This was reflected in the 1980 Industry Act which obliged it to promote private ownership of industry by disposing of its own industrial investments. The main strategic objective, however, remains: 'Indeed what will distinguish the Agency from many other development agencies will be the potential for an integrative approach towards the industrial and urban problems which exist in Scotland' (Cunningham 1978).

The agency's board is appointed by the Secretary of State for Scotland. Its membership is drawn from industry, finance, local government, and trade unions. All members serve in a part-time capacity except for the Chief Executive who is a salaried employee of the agency.

The SDA took over three bodies and 470 staff. The most significant was the Scottish Industrial Estates Commission which administered some 2.8 million square metres of factory space. The agency thus became the largest single industrial landlord in Scotland. It also absorbed the Small Industries Council for Rural Areas of Scotland which formed the nucleus of the agency's small-business division. Finally, it also took over the administration of land clearance grants to local authorities from the Scottish Office.

The SDA is formally accountable to Parliament through the Secretary of State for Scotland, who is empowered to issue directives and, in conjunction with the Treasury, sets the agency's financial framework. Within the Industry Department for Scotland there is a division handling general relations with the agency. At the same time, the SDA has frequent direct contacts with other parts of the department with responsibility for regional and sectoral policy and with the Scottish Development Department over urban renewal projects.

Within the agency there are a number of functional divisions covering its main activities, like property management, industrial policy, area development, and inward investment. These divisions are supported by strategic policy and administrative divisions. Significant restructuring of the organization has occurred since the appointment of the agency's second Chief Executive, George Mathewson, in 1981. Responsibilities for urban renewal and special area projects like GEAR were centralized into an Area Development Directorate. Strategic planning of both sector and area policies was combined with industry services and evaluation within a Planning and Projects Directorate. The Financial Directorate has assumed responsibility for managing the agency's physical assets, although the Property and Environment Directorate continues to administer the agency's industrial estates but is also increasingly involved in commercial developments such as the St Enoch project in Glasgow's city centre.

In 1983, LIS, a new agency jointly sponsored with the Scottish Office, was created to attract new inward investment. This organizational innovation could be seen as a compromise over who should be responsible for this activity. Because LIS is responsible to both the agency and Scottish Office, it has benefited from its limited relationship with both sponsors and has successfully managed to operate with a large measure of independence.

Following changes in the agency's industrial investment guide-lines in 1981, a new wholly owned subsidiary company called Scottish Development Finance was set up to oversee the SDA's activities in this area. This body has brought in private sector interests, thereby subjecting the agency's strategy to external commercial appraisal and giving this activity greater political legitimacy.

Organizational futures As the SDA has become more specialized and experienced in its various roles, it has developed in ways that could not be foreseen in 1975. Similarly, in the next ten years we might expect further changes in both task and structure. There are two possible futures which could significantly alter the present role of the agency and its place in the Scottish economic policy community. The first is what we call the *concertation* model in which the agency would enhance its role with greater powers and resources.

It can be argued that currently the agency is unbalanced. It has responsibilities for the infrastructural needs of industry through its factory development and estates management role, and also the responsibility to invest, attract, and advise businesses. However, it has no control over the third vital element of economic regeneration—namely, the provision of skilled labour. Responsibility for labour planning and training rests with the MSC operating within a UK context. While national programmes should continue to be run by the MSC, it is legitimate to argue that there are particular needs which might be better provided by a distinctively Scottish agency with responsibility for economic development.

Such a reorganization would have problems—not least, opposition from the MSC—but the SDA is already involved in small-scale training initiatives (for example, in GEAR). If there are special needs in Scotland at the local level, which the MSC cannot effectively meet because of its national orientation, then the development of a more systematic training role for the SDA might be forced onto the policy agenda.

The major advantages of the concertation model thus include the location of economic development functions in an agency outside the immediate governmental system and with a capacity to take a long-term view of the needs of the Scottish economy. This responsibility would need to be matched by the concentration of powers necessary to perform these tasks. At the same time, a regional agency would have to develop close links with other local bodies in order to harmonize and co-ordinate activities. A regional focus could also enhance the opportunity for bending resources and leverage at local levels.

There are disadvantages, however. Firstly, it is questionable how much autonomy such an agency would enjoy in its relationship with central government, especially if its own priorities came into conflict with macro-economic policies. Secondly, such an agency could well face hostility from local authorities unless it engaged in a consensus or bargaining relationship. There would also be a need to work out a satisfactory relationship with the MSC addressing the feasibility of separating regional labour and training policy from a UK context. Thus, any moves towards the concertation model require very careful consideration of the political and economic consequences.

An alternative to the above is the *fragmentation* model which builds on the increasing specialization of the agency to create semi-autonomous operational units within the framework of a holding company. Such an organization might comprise an Investment Bank providing funds to Scottish enterprise on a short- and long-term basis, and an Area Development Agency with responsibility to regenerate designated areas combining crisis management with strategic renewal as outlined previously. The theoretical advantage of such an organizational structure is that it would combine clearly identified overall objectives with the maximum degree of operational specialization, essentially addressing the classical organizational problem of marrying integration and differentiation (Lawrence and Lorsh 1967). The obvious disadvantages are that specialization can lead to fragmentation and loss of strategic overview and that this is an artificial division which does not correspond coherently with the nature of the problems faced. Fragmentation might also increase problems of internal co-ordination and conflict management. Again, such reorganization would require careful consideration. At present, there would not seem to be any great advantage in pursuing such an option.

Conclusion

Clydeside, despite the many policies and agencies devised to assist in its economic regeneration, remains one of Britain's most depressed areas. While the visitor may now be impressed by the massive improvement in the physical environment, by the revitalization of the central business district, and by its acquisition of major hotels and retail and conference developments, a more wide-ranging survey will reveal the very high rates of unemployment tucked away on the peripheral housing schemes. We have argued elsewhere (Lever and Moore 1985) that only within a programme of national economic regeneration is substantial relief likely to be brought to cities like Clydeside. However, this is not to say that the policies described here have been unsuccessful, for it is difficult, if not impossible, to estimate what the position would have been had there been no such intervention.

Back-of-envelope calculations indicate that about 20 000–30 000 jobs were created or maintained in the period 1980–4 in the conurbation. Allowing for multiplier effects, this number might approximately be doubled to 40 000–60 000 jobs. This figure should be placed in the context of registered unemployment of close to 200 000. By way of conclusion, we can only suggest that, without the exceptionally high levels of intervention into the economy experienced by the Clydeside conurbation, its employment position in mid-1985 would have been significantly worse.

References

Bull, P. (1981), 'Redevelopment Schemes and Manufacturing Activity in Glasgow', *Environment and Planning*, 13, 991–1000.

Cameron, G.C. (1979), 'The National Industrial Strategy and Regional Policy', in D. Maclennan and J. B. Parr (eds.), *Regional Policy: Past Experience and New Directions*, Oxford: Martin Robertson.

Community Business Scotland (1981), *The Development of Community Business in Scotland*, Glasgow: Community Business Scotland.

Cunningham, E. (1978), *Regional Policy: The Role of a Development Agency*, Glasgow: SDA.

Department of Industry (1983), *Regional Industrial Development*, Cmnd 9111, London: HMSO.

Fothergill, S., and Gudgin, G. (1982), *Unequal Growth: Urban and Regional Employment Change in the United Kingdom*, London: Heinemann.

—— Kitson, M., and Monk, S. (1982), 'The Profitability of Manufacturing Industry in the UK Conurbations', *Working Paper* No. 2, University of Cambridge, Department of Land Economy.

GDC (1978), *Council Minutes*, Print 7, pp. 619–20, Oct., Glasgow District Council.

GDC–SRC (1985), *Proposed Joint Initiatives in Drumchapel and Greater Easterhouse*, Report by Chief Executive to Policy and Resources Committee, Glasgow: Glasgow District Council.

Hayton, K. (1983), 'Employment Creation in Deprived Areas: The Local Authority Role in Promoting Community Business', *Local Government Studies*, 9(6), 39–55.

Henderson, R.A. (1980), 'The Location of Immigrant Industry within a UK Assisted Area: The Scottish Experience', *Progress in Planning*, 14(2), 103–226.

Hood, N. (1984), 'The Small Firm Sector', in N. Hood and S. Young (eds.), *Industry, Policy and the Scottish Economy*, Edinburgh: Edinburgh University Press.

—— Milner, M., and Young, S. (1982), *Growth and Development in Small Successful Manufacturing Firms in Scotland*, Report prepared for the Scottish Economic Planning Department, Edinburgh.

Hughes, J. (1981), *Britain in Crisis*, Nottingham: Spokesman.

Lawrence, P.R., and Lorsh, J.W. (1967), *Organisation and the Environment: Managing Differentiation as Integration*. Boston, Mass.: Harvard Business School.

Lever, W.F. (1986), 'Labour and Capital', in W. F. Lever (ed.), *Industrial Change in the United Kingdom*, London: Longman.

—— and Moore, C. (1985), 'Future Directions for Urban Policy', in W. F. Lever and C. Moore (eds.), *The City in Transition: Policies and Agencies for the Economic Regeneration of Clydeside*. Oxford: Clarendon Press.

McArthur, A.A. (1984), 'The Community Business Movement in Scotland: Contributions, Public Sector Responses and Possibilities', *Discussion Paper* 17, Centre for Urban and Regional Research, University of Glasgow.

McGregor, A. (1978), 'Unemployment Durations and Re-employment Probability', *Economic Journal*, 88, 693–706.

—— (1983), 'Neighbourhood Influences on Job Search and Job Finding Methods', *Business Journal of Industrial Relations*, 21, 91–9.

McKean, R. (1975), 'The Impact of Comprehensive Development Area Policies on Industry in Glasgow', *Urban and Regional Discussion Papers* 15, University of Glasgow.

Marquand, J. (1980), 'Measuring the Effects and Costs of Regional Incentives', *Working Paper* 32, Government Economic Service, Civil Service College, London.

Mason, C. (1986), 'The Small Firm Sector, in W. F. Lever (ed.), *Industrial Change in the United Kingdom*, London: Longman.

Moore, C. and Booth, S. (1984), 'Urban Economic Adjustment and Regeneration: The Role of the Scottish Development Agency', *Inner City in Context Working Paper* No. 10, University of Glasgow.

Storey, D. (1982), *Entrepreneurship and the New Firm*, London: Croom Helm.

Strathclyde Regional Council (1984), *Strathclyde Economic Trends*, No. 6, Glasgow: Strathclyde Regional Council.

Townsend, A.R. (1983), *The Impact of Recession*, London: Croom Helm.

Tyler, P., Moore, B., and Rhodes, J. (1980), 'New Developments in the Evaluation of Regional Policy', Paper presented to the ESRC Urban Economics Seminar Group, University of Birmingham.

—— —— —— (1984), 'Geographical Variation in Industrial Costs, *Discussion Paper* 12, University of Cambridge, Department of Land Economy.

Young, S. (1984), 'The Foreign-owned Manufacturing Sector', in N. Hood and S. Young (eds.), *Industry, Policy and the Scottish Economy*, Edinburgh: Edinburgh University Press.

6

The West Midlands: An Economy in Crisis

*Ken Spencer**

The West Midlands was selected by the Economic and Social Research Council (ESRC) as an example of an area which had traditionally been regarded as prosperous and buoyant but which had more recently suffered a rapid decline in industrial prosperity and job prospects. This decline has been particularly acute in terms of the manufacturing base of the industrial heartland of the nation. Paradoxically, the area's strong industrial base, upon which the prosperity and progress were founded, was itself the cause of decline. However, it is clear that manufacturing decline was already evident before the major recession of 1979 and subsequently. The symptoms of decline were apparent much earlier, but gained little recognition as such outside the area itself.

Given this very rapid rate of decline in manufacturing industry, public authorities have responded with a great variety of innovative schemes and policies to combat problems. At the same time, the large manufacturing industries, including vehicles and components manufacturers, have had to struggle hard to survive. In so surviving, many companies are now very different from what they were when initially faced with crisis.

The project has concentrated upon the West Midlands County Council area, but has extended, as appropriate, both to the nearer parts of the region on the county periphery and to the wider West Midlands Region. Within the ESRC project framework, the research of the West Midlands' team has been concentrated on four key areas. First, an examination of the nature of change in the West Midlands economy since 1960 has been related to changes in local, national, and international economic environments. Second, the role of public policy in terms of actions and impacts on the West Midlands has been considered. Third, because of the heavy dependence of the local economy upon large manufacturing firms, we have paid particular attention to de-

* This chapter provides a summary of much of the material which has been produced by the West Midlands research team for the ESRC Inner Cities Research Programme. The research team was drawn from the Centre for Urban and Regional Studies and the Institute of Local Government Studies in the University of Birmingham. The team members on whose contributions this chapter is based were Ken Spencer (co-ordinator), Richard Batley, Norman Flynn, Valerie Karn (until Easter 1984), John Mawson, Barbara Smith, and Andy Taylor. The work of the team, largely through a series of nineteen Working Papers and six Working Notes, is gratefully acknowledged.

industrialization and corporate strategy as exemplified by the twenty-six largest firms. Fourth, we have been concerned with the inner city in context, not simply with the inner city *per se*. It is this wider context that has led to the situations that currently exist in the inner urban areas of the West Midlands.

Historically, the West Midlands has the reputation of being a prosperous region successfully adapting to industrial change over the years. This prosperity was founded upon the core industries of coal, steel, and metalworking. With the decline of these basic industries, the engineering industries, utilizing the skills of the metalworkers, were developed. The suppliers of these major industries were developed locally and the West Midlands emerged as the main centre of manufacturing industry in Britain—the industrial heartland.

The West Midlands economy adjusted successfully to the recessions of 1870–80 and 1920–30 and to changes in demand after each World War. New growth sectors had emerged and the area's interdependence led to a diffusion of the benefits of growth. However, the conditions which were to make the economy vulnerable in the 1970s were already accumulating. Industry had become larger scale, less diverse, and less locally interdependent. There was increasing concentration on vehicle manufacture and engineering. This less flexible economic structure had come to depend on a prosperous domestic and Commonwealth market and on low levels of international competition. The conditions were also growing for firms to choose to expand outside the county and, by the 1950s, government policy was encouraging such dispersal and discouraging increasing investment (Batley 1984).

The prosperity of the post-1945 West Midlands economy disguised a vulnerability which its earlier heterogeneity had avoided. A high-wage economy was built on the prosperity of the leading sectors of the economy and on the effect of full employment with semi-skilled and unskilled labour benefiting in wages and security of employment from labour scarcity and mass unionism (see also Wood 1976; Liggins 1977).

The economy was buoyant. In the 1930s depression, Birmingham and Coventry unemployment levels were lower than the national average. Up to the mid-1960s the West Midlands was one of the fastest-growing regions, with high wage levels and high economic activity rates leading to family incomes above those in all other regions except the South East. The foundation was a seemingly strong manufacturing base in vehicles, metal manufacture, and engineering.

By the 1980s, the region was suffering unemployment levels exceeding those in many of the traditional development areas. The West Midlands had shifted from prosperity and buoyancy to a state of rapid decline and crisis—a situation brought about by a collapse of its strong but poorly diversified manufacturing base (Mawson and Taylor 1983).

By December 1984, unemployment in each of the districts of West Midlands County was: Birmingham, 20.5 per cent; Sandwell, 19.4 per cent;

Wolverhampton, 19.8 per cent; Coventry, 18.8 per cent; Walsall, 16.2 per cent; Dudley, 14.4 per cent; and Solihull, 11.9 per cent; while the county figure was 18.3 per cent. For the Birmingham Inner City Partnership area the rate was 23.7 per cent.[1] By 1984, one in three households in the county lived at or below the official poverty line, while the numbers claiming supplementary benefit had risen 50 per cent since 1980.

Industrial Decline

In the period 1980–3, total employment in the region fell by 16 per cent, while manufacturing employment declined by over 24 per cent (December 1979 to June 1982, a loss of 230 000 jobs in just 30 months). The previous increase in service employment, which to some small extent had eased overall unemployment, levelled off. Job loss, especially in manufacturing industry, has continued. Of the notified 155 578 redundancies in West Midlands County between 1 January 1981 and 31 August 1983, some 82.2 per cent were in manufacturing (13.5 per cent were in services and 4.3 per cent others). Of those notified redundancies in manufacturing industry, just over 85 per cent were in five key sectors. Vehicles represented 25.3 per cent, metal goods (not elsewhere specified) 18.7 per cent, metal manufacture 15.7 per cent, mechanical engineering 14.6 per cent, and electrical engineering 10 per cent. Notified redundancy figures should be treated with caution, but they do indicate the continuing rapid employment decline in the key sectors of the West Midlands economy (Taylor 1983).

The competitive position of many of these industries is poor. The process of de-industrialization of the British economy is taking a very heavy toll on the West Midlands, particularly within the West Midlands County area. More specifically, job losses are particularly acute in the inner city wards of Birmingham, Sandwell, Wolverhampton, Coventry, and Walsall. The effects of this dramatic decline in employment opportunities are most strongly felt within these inner city wards, many of which have a high incidence of ethnic minority residents, especially in Birmingham and Wolverhampton (Taylor 1984). In Birmingham's Inner City Partnership core area in 1981, 43.1 per cent of residents (113 314 people) were in households where the head was born in the New Commonwealth or Pakistan. In the Partnership area itself (covering 65 per cent of the city's population compared with 27 per cent in the core area), there were 21.3 per cent of such ethnic minority residents.

In comparing economic performance indicators—investment, output, gross domestic product (GDP)—the region has scored badly, with a rapid deterioration in recent years. At the same time, the region suffers from a large negative balance of government expenditure, the difference between expenditure and tax receipts in the region being much greater than in the relatively prosperous South East which is in net surplus (Mawson and Taylor 1983).

In 1978, 48.6 per cent of employment in West Midlands County was in

Table 6.1. *Broad Industrial Structure of West Midlands County: Percentage of Employees by Sector, 1961, 1971, 1978, 1981*

Sector	1961		1971		1978		1981	
	WMC	GB	WMC	GB	WMC	GB	WMC	GB
Agriculture	0.28	2.55	0.16	1.94	0.16	1.67	0.17	1.75
Mining	0.49	3.28	0.26	1.82	0.22	1.58	0.19	1.58
Manufacturing	64.49	39.31	54.46	36.42	48.61	31.95	42.30	28.01
Construction	4.92	6.61	4.50	5.65	4.52	5.30	4.46	5.15
Services	29.73	48.26	40.42	54.18	46.49	59.29	51.24	63.51

Note: WMC, West Midlands County; GB, Great Britain.

Source: Department of Employment, Annual Census of Employment.

manufacturing industry, compared with 32.1 per cent nationally, while the service sector in the area accounted for 46.5 per cent, compared with 52.6 per cent nationally. Four key sectors accounted for 70.6 per cent of total manufacturing employment in 1978 (37.4 per cent nationally). These were: metal manufacture, 14.6 per cent; mechanical engineering, 11.5 per cent; vehicles, 23.9 per cent; metal goods, 20.6 per cent (Taylor 1983). Table 6.1 indicates this changing employment structure.

Thus the picture is of a local economy heavily dependent upon a few key industries. Prosperity was built upon this, but these same foundations have led to economic collapse in a period of recession. 1966 was a bad year for the car industry, and its 'knock-on' effects were felt much later (Compton and Penketh 1977). The 1971–3 slump also badly affected the area (Taylor 1983). The spread of this decline has been compounded by the high level of interlinkages and interdependence between a small number of large firms and a large number of small firms. This level of interlinkage has been greater in the West Midlands region than in any other region, and this is especially the case in West Midlands County. As a result, the industrial base has been more vulnerable to external economic forces. The decline has been swiftly transmitted through the area's economy and has been heavily concentrated in the dominant sectors of the manufacturing economy. Areas outside the county have also been affected, including the New Towns of Telford and Redditch—particularly the former.

At region and county levels, the West Midlands is characterized by low levels of net output, a poor investment record, and low levels of productivity. The area is outstandingly poor in its record of industrial investment across all manufacturing industries. Old deteriorating premises and lack of space for expansion have been particularly acute in the older industrial areas where premises were constructed mainly during the nineteenth century.

This issue of land and premises is prominent in the urban-to-rural industrial shift argument as put forward by Fothergill and Gudgin (1979, 1982) and Fothergill *et al.* (1984, 1985). The argument turns on their trends of declining numbers of workers per unit of factory space and upon industry's increasing demand for further space (Fothergill *et al.* 1985). Thus space constraints are the key cause of city employment decline. An analysis of land and premises in the West Midlands shows that land supply is limited and choice constrained in the county (West Midlands Forum of County Councils 1985; CURS 1984; West Midlands County Council 1983; City of Birmingham 1984). In terms of the quality of vacant industrial premises, some one-third were unusable (West Midlands Forum of County Councils 1985), while some 74 per cent of establishments in the Birmingham inner city core area operated on more than one storey, compared with only 25 per cent in rural areas (Fothergill *et al.* 1985, 26).

Thus, for Fothergill *et al.*, the higher price and inadequacy (in quantity, choice, and quality) of industrial land and premises, combined with floor-space pressures, is the key to employment loss in West Midlands County and the shift of employment to places where constraints are less. This is Keeble's 'constrained location theory' (Keeble *et al.* 1983, 406–7). Our own results would lead to the conclusion that the supply of land and buildings sets a ceiling for existing industry and influences its performance to a degree, but rather as a hindrance, not as an overwhelming force (Smith 1985). Such emphasis on land and buildings *per se* ignores the weakness of external investment through new firms and branch plants, loss of markets, and lack of diversification in existing firms where management, design, quality of product, and marketing seem strong factors unaffected by premises. Land and premises do not explain the corporate restructuring processes of the largest twenty-six firms in the West Midlands, whose impact on the local economy has been particularly significant (see later). Nor do such approaches explain the closure of newly built factories with modern equipment.

In the early 1960s, GDP per head of regional population was nearly 10 per cent higher than the United Kingdom (UK) average, being second only to that of the South East. This position deteriorated from the mid-1960s. Since 1976, GDP relative decline has accelerated, with GDP falling from 98.1 per cent of the UK average in 1976 to 90.6 per cent in 1981. Thus, from being second to the South East in 1976, by 1981 the West Midlands was the lowest of all English regions (only Northern Ireland and Wales were lower).

The West Midlands has not seen expansion through growth industries. Indeed, employment in many of the growth industries identified by the West Midlands Economic Planning Council (1971) has declined in the county. This may suggest that the search for, and encouragement of, growth industry may be more difficult for the West Midlands than at first appears. This may be due to industrialists' image of the county—many areas of poor industrial environment—and their general attitude to residential environments and work-force characteristics (Cochrane 1984).

Poor employment performance cannot be explained solely by reference to

Table 6.2. *West Midland County: Industrial Sectors with a Location Quotient Greater than 2 in 1981*

Rank	MLH	Location quotient
1	Bolts, nuts, screws	6.8
2	Copper, brass and other copper alloys	5.5
3	Steel tubes	4.3
4	Jewellery and precious metal	3.7
5	Metal industries n.e.s.	3.3
5	Iron castings	3.3
7	Metalworking machinery	2.9
8	Motor vehicle manufacture	2.8
9	Aluminium and aluminium alloys	2.6
10	Other electrical goods	2.3
11	Telegraph and telephone apparatus and equipment	2.2

Source: As Table 6.1.

the West Midlands' adverse industrial structure. Shift–share analysis (for an outline of this technique see Danson *et al.* 1980) shows a large proportion of employment decline in the county being due to factors other than purely national trends or structural mix. Economic processes influencing change within key sectors and key companies are important. Many companies have invested elsewhere in Britain or abroad; some have switched products; others have moved from production to distribution of imported goods; others are located on confined sites, often with poor access and infrastructure; some suffer poor-quality, often worn-out, premises (often through lack of investment); while others have poor technology and high unit production costs.

The level of import penetration measured by the ratio of imports to home demand in the five key sectors of the West Midlands economy reflects the failure to compete in specific areas. From 1973 to 1982, import penetration in metal manufacturing rose by 48 per cent, in mechanical engineering by 38 per cent, in electrical engineering by 81 per cent, in vehicles by 96 per cent, and in metal goods by 60 per cent. This has to be balanced against exports. One measure is the ratio of exports to manufacturers' sales. These export/sales ratios have increased over the same period by 60 per cent in metal manufacture, 33 per cent in mechanical engineering, 80 per cent in electrical engineering, 28 per cent in vehicles, and 31 per cent in metal goods. It is these last two sectors where the shifting trade balance has resulted from failure to compete effectively (Flynn and Taylor 1984). In 1978, these two sectors accounted for 36 per cent of manufacturing employment in the West Midlands. These are the sectors which are heavily represented in the West Midlands economy. Table 6.2 shows location quotients for the West Midlands County. A location quotient of 2 means the county has twice the national level of employment within that sector.

Thus the regional economy was in poor shape to cope with a deep recession. Markets had been progressively lost, and profits and investment had been low, leading to relatively high unit costs and an inability to develop new products for existing or new markets. This led to further decline in profitability. New investment was attracted away from the West Midlands County to outlying towns beyond the Green Belt and to other areas where investment incentives were available through regional policy (Bentley and Mawson 1985a), and European Economic Community (EEC) funds. (Mawson *et al* 1983). Telford and Redditch, as well as other areas which were not town areas, attracted industry from the conurbation area.

Massey and Meegan's study (1982) of company response to recession identified three forms of production reorganization: intensification, keeping the same production process but speeding up the work; investment and technological change; rationalization to reduce capacity in the face of reducing demand. It is possible to identify other responses to changing market circumstances:

(*a*) Pay can be frozen or reduced in the negotiation stage, by cheaper new labour, by reducing overtime and shift work
(*b*) Components and materials can be bought from cheaper suppliers which may be overseas
(*c*) Companies may switch to new products
(*d*) Products can be improved
(*e*) Sales and marketing effort can be improved
(*f*) New markets can be opened as old ones disappear
(*g*) Production can shift to a new location where costs are lower
(*h*) Companies may go out of business as competition eliminates those with poorer asset/liabilities ratios
(*i*) Companies may liquidate and be taken over by others wishing to keep plants operating, but only if asset values are reduced through liquidation
(*j*) Companies may retain capacity, but reduce output temporarily in the face of perceived temporary reduction in demand.

The Confederation of British Industry's (CBI's) survey of local industrialists shows the last response to changing markets to be widespread in the West Midlands (CBI 1983). Gaffikin and Nickson (1984) have shown that many of the larger companies in the area are transferring assets to other locations, often outside Britain. This is reinforced by our analysis of the top twenty-six companies in the local economy and their response to survival and development strategies, (Flynn and Taylor 1984).

The larger firms which had diversified their products from the 1960s onwards were in a better position to cope with recession. GKN shifted from manufacturing fasteners to importing and distributing them. The recession has accelerated the processes of change and forced many companies to cut down their operations in the West Midlands. Those which had not already diversified by the late 1970s were more likely to run into financial and cash-flow difficulties—for example, Rubery Owen, Duport, and Dunlop.

Manufacturing employment in the West Midlands economy has been declining since 1965. Between 1965 and 1981 some 369 500 jobs in manufacturing were lost. Since 1971, the manufacturing base has contracted by over 34 per cent. The vehicle industry alone lost over 82 000 jobs in the ten years to 1981—a loss of over 40 per cent. Similarly, over 40 per cent of the jobs in metal manufacturing went between 1971 and 1981.

At the same time, employment performance in the service sector in the county has been poor. Whereas between 1971 and 1978 service employment increased nationally by almost 21 per cent, the county increase was only some 10 per cent (on a low base). Between 1978 and 1981 service employment decreased by nearly 3 per cent, in contrast to a national growth of just under 2 per cent. Service sector employment has fallen well short of matching the decline in manufacturing jobs. Furthermore, during the period of service sector growth, the jobs lost in manufacturing were mainly full-time male, while those gained in the service sector were mainly part-time female (Smith 1984a,d).

To summarize: the traditional manufacturing sectors all shed employment throughout the 1970s, and the present recession has merely accelerated a longer period of decline. Many of the companies in the local economy have diversified. They have not necessarily located the production of the new products in the West Midlands. The local county economy has not diversified out of metal; nor has it made the shift from vehicle manufacture into new industrial technologies or products, nor to any degree into the service sector. Paradoxically, the local economy is now more highly specialized in its manufacturing base than in 1971.

Unemployment

As already stated, at December 1984 unemployment in West Midlands County was 18.3 per cent. It is also clear that the present high number of people without work is not a recent cyclical phenomenon, but is part of a longer-term trend of worsening levels of those without work which can be traced back to the mid-1960s (Taylor 1984). Unemployment in West Midlands County has increased relative to the national average via a 'ratchet effect' of significant upward shifts in unemployment following economic downturns without subsequent gains of employment to compensate. In the mid-1960s, unemployment rates in the West Midlands were about half of the national figure. The economic downturns of 1971–72 and 1975–6 saw the area rapidly suffering the effects of recession; however, with national recovery, employment did not return at the national rate. Hence the area entered each new downturn from a higher unemployment base. Until the late 1970s, unemployment was lower in Dudley–Sandwell and Walsall than in the county, but it has rapidly increased in these areas in the 1980s. Thus the recession in the Black Country was reflected in unemployment growth later than in Birmingham, while Coventry suffered to a greater extent in the earlier 1970s than other areas within the county (Smith 1984d).

At October 1983 the pattern of unemployment analysed by wards indicated that Birmingham had three inner city wards with over 40 per cent unemployment, while seven other wards were recorded with over 30 per cent unemployment in Birmingham, with two in Coventry and one in Sandwell. Inner city wards were by far the most vulnerable. These high unemployment levels are accompanied by a marked increase in the number of long-term unemployed. Long-term unemployment is a very significant criterion of relative economic stress and its incidence is changing the map of job opportunities in Britain (Ball, 1983; Green 1984).

Ball (1983) has highlighted this for the West Midlands Region which in 1974 had the second-lowest UK regional proportion of persons unemployed for over one year, at 21.2 per cent. By 1982, 38.6 per cent were long-term unemployed, and by July 1984 nearly 50 per cent of the region's unemployed (compared with just over one-third nationally) were long-term unemployed. This is the highest long-term unemployment figure of all UK regions. In human terms, this represents 158 800 persons, of whom 94 000 had been without work for more than two years. Long-term unemployment, which regionally saw 18 per cent of the unemployed being unemployed for three years or more at July 1985 (the highest regional figure), is geographically concentrated within the metropolitan county, especially within Birmingham, Coventry, and Wolverhampton, and especially within the core area of the Birmingham Inner City Partnership zone.

Between 1971 and 1983 the number of employees in employment in the Partnership zone (core and outer area) saw a decrease of 46 per cent in manufacturing (−40 per cent for Birmingham and −32 per cent for Great Britain), an increase of 4 per cent in service employment (no change for Birmingham and +11 per cent for Great Britain), and a decrease of 41 per cent in construction (−3 per cent for Birmingham and −19 per cent for Great Britain). The loss of construction industry jobs has particularly hit the inner city (Birmingham Inner City Partnership 1984, 2). At the same time, inner city core area residents in work in 1981 were more dependent upon buses (43 per cent) and foot (19 per cent) as a means of travel to work than residents in the city as a whole (34 per cent and 13 per cent respectively). Only 28 per cent used a car to travel to work, compared with 45 per cent in Birmingham: thus inner city residents did not possess the same degree of implied flexibility of work location through journey to work by car.

The inner city socio-economic structure is also skewed to those occupations more likely to suffer as a result of employment structure contractions. The inner city core area of Birmingham provides an ethnic minority dimension to the unemployment pattern. In 1981, 43.1 per cent of the population in the Partnership core area were in households where the head was born in the New Commonwealth or Pakistan (compared with 21.3 per cent for the Partnership area as a whole, 15.2 per cent for Birmingham, and 4.5 per cent for Great Britain).

These groups are heavily concentrated both within Birmingham and within its inner city core area. Unemployment among these ethnic minority groups is more than 50 per cent higher than the equivalent white rates within the Partnership area. Within Bangladeshi households the rate is over 100 per cent higher, closely followed by high rates in Pakistani households (1984, Birmingham Inner City Partnership). There is thus a much more acute unemployment problem among these ethnic minority groups within the inner core area of Birmingham. In Sandwell and Wolverhampton, and to a lesser extent in Walsall and Coventry, this same strong relationship of inner city unemployment and ethnic minorities is found. In August 1982, 54.4 per cent of those of ethnic minority origin registered unemployed within the West Midlands County were to be found in Birmingham. Together with racial discrimination in job allocation, this means that such groups face extreme disadvantage in obtaining work (West Midlands County Council 1983, paras. 3.1.22 and 3.1.24).

Typically, the level of unemployment among women is lower than that for men. However, this hides the fact that unemployment was growing faster among women than among men during 1970s. Since 1980, however, the rate of increase in male unemployment has generally exceeded that of females owing to the post-1979 recession and its particular impact upon the manufacturing sector which is dominated by male employment. (Taylor 1984, 47–8).

All the West Midlands County travel-to-work areas are to be found within the top ten nationally of rates of long-term unemployment (over one year) among travel-to-work areas. At January 1984, Liverpool was top with 50.3 per cent of its unemployed out of work for over one year, second was Wolverhampton at 48.7 per cent, third Birmingham at 47.6 per cent (though with a much higher absolute number of persons involved), fourth was Dudley–Sandwell at 46.9 per cent, sixth Walsall at 45.6 per cent, and tenth Coventry at 44.1 per cent (Birmingham Chamber of Industry and Commerce, 1984, Table 2).

One might expect that higher unemployment would lead to more people becoming self-employed. Our evidence indicates no relationship between levels of unemployment and levels of self-employment. Thus self-employment change does not indicate any resurgence of small business in the region (Smith 1984c).

Relative Growth and Decline in Population and Employment

Between 1961 and 1981 the population of West Midlands County fell by 3 per cent, compared with a rise of 23 per cent in the rest of the region. The overall rise in population of the region was 8 per cent—higher than the national rate of 6 per cent for England and Wales. In the travel-to-work areas overlapping the county boundary there was a 4.7 per cent increase compared with a 15.5 per cent increase in the rest of the region (Smith 1983).

The main fall in residential population was in Birmingham, which lost 15 per cent of its population; Sandwell lost 9.3 per cent; Wolverhampton lost 2.7 per cent; and Coventry lost 1.4 per cent. By contrast, Walsall gained 8.2 per cent, Dudley 17.9 per cent, and Solihull 55.4 per cent (mainly owing to a Birmingham overspill estate at Chelmsley Wood). Birmingham's population loss occurred both in the 1960s and 1970s, with only a little extra loss in the 1970s compared to the 1960s. Sandwell was the only other area to lose population in the 1960s, but the fall was far below that of the 1970s. In the 1970s, decline was much more general, affecting all the metropolitan districts except Solihull and Dudley. Birmingham fell into decline earliest and fastest. Between 1961 and 1981 the population shift redistributed 6 per cent of the region's population from the declining to the growing parts of the region.

Over the same period 1961–81 it is possible to compare population characteristic changes in terms of the numbers of economically active population, the employed population, and unemployment rates. (Smith 1984d). By this means it is possible to measure areas of relative growth and decline. Table 6.3 sets out these data. The heart of decline lies in Birmingham. In Birmingham and Solihull (but primarily in Birmingham) there was a reduction of 14 per cent in the economically active population, a reduction of 24 per cent in the employed population, and 617 per cent increase in unemployment (much higher in Birmingham than Solihull). The Black Country authorities (Walsall, Dudley, Sandwell, and Wolverhampton) hardly lost economically active population, but lost 15 per cent of their employed persons and experienced a more severe rate of unemployment increase at 732 per cent. The situation in Coventry closely parallels that of the Black Country. Growth was located outside the metropolitan county area. The fringe of the West Midlands County area was the area of highest growth with a 48 per cent increase in the economically active population and a 37 per cent increase in the employed population, though unemployment still rose by 489 per cent. The rest of the West Midlands region also benefited from growth (that is, the area outside the West Midlands County's travel-to-work areas). This zone saw a 14 per cent rise in economically active population and 6 per cent rise in the numbers of employed people. It also had a lower growth rate in unemployment at 391 per cent (though still above the figure for England and Wales at 369 per cent). In England and Wales there had been a 6 per cent rise in the economically active population and a 1.5 per cent increase in the number of people employed during the period 1961–81.

Thus the main locus of growth, which was also accompanied by unemployment rates relatively lower than most areas of the region (though still higher than the national average), was in those areas bordering the West

Table 6.3. *Percentage Changes in Population and Employment in the West Midlands, 1961–1981*

Area	Population	Economically active persons	Employed persons	Unemployment rate increase
Birmingham–Solihull	−8.1	−14	−24	+617
WMC	−3.0	−8	−20	+676
Black Country	+2.5	−3	−15	+732
Coventry	−1.4	−2	−15	+724
WMC TTWA	+4.7	−0.3	−12	+623
WMR	+8.2	+4	−6	+557
WMR excl. WMC TTWA	+15.5	+14	+6	+391
WMR excl. WMC	+23.4	+22	+13	+414
Outer fringe of WMC TTWA and WMC	+49.9	+48	+37	+489
England and Wales	+6.1	+6	+1.5	+369

Notes: WMC, West Midlands County; WMR, West Midlands Region; TTWA, travel-to-work area.

Source: Smith (1983, 1984d).

Midlands County boundary, beyond the Green Belt in the newer expanding communities. All this supports the urban-to-rural shift hypothesis. It can be argued therefore that, rather than there being population decline in the West Midlands County, a more embracing concept is that population dispersal to a widening city region is taking place.

However, that population dispersal has been highly selective, with the result that social polarization in Birmingham is extremely marked. This contrast between the inner area and outer areas, on economic, social, and housing indices, is also more marked than in any other Inner City Partnership area nationally (Department of the Environment, Inner Cities Directorate, 1983, 4). This particularly reflects ethnic minority group concentration in the inner city. The inner/outer gradient is especially steep in Birmingham where the degree of socio-economic polarization is greater than anywhere else in the county except Teesside (Begg and Eversley 1986). At the same time the Birmingham inner city area is the most disadvantaged of all Inner City Partnership areas in the country. It is the dominant ethnicity factor with its strong correlation with high unemployment, low socio-economic structure, and poor educational attainment which is the cause of this strong polarization. Much of the movement out of the inner city has been of white people, leaving behind housing taken up by second-generation ethnic minority residents. Strong ethnic cultural ties in the inner city and their attraction to ethnic minorities has led to the retention of minority groups in the inner city. This is reflected in the environment—for example, food shops, butchers, mosques, temples, banks, travel firms, clothing shops, and so on.

Those who can afford better housing are moving out and, with reducing public and private sector housing investment in the inner city, many of those in greatest housing need will increasingly live in environmental and housing conditions which are deteriorating (West Midlands Forum of County Councils 1982). The racial polarization of the inner city owner-occupied housing market means that Asians and West Indians (to a lesser extent) bear the brunt of problems—notably loan problems and maintenance and improvement costs (Karn 1983). Housing market mechanisms and the operation of the public sector are, because of reduced public and private resources, accelerating the spiral of decline in the inner city areas. Many will be trapped in poor housing conditions and, with the lack of employment opportunities, will find it difficult to gain access to better housing and environmental conditions (Johnson 1982, 198). Meanwhile inner city housing conditions continue to deteriorate.

Investment in other services provided by local authorities and other public agencies has often been subject to cut-back; in many cases this has led to an erosion of service quantity and quality, especially within the inner urban areas. There is no mechanism at regional or city level for monitoring the effect of declining investment on urban services and upon individuals, social groups, and households; yet the evidence would suggest this is particularly vital (Stewart 1981).

Earnings

Relative earnings in the West Midlands fell over the period 1970–83. The fall in earnings was from top regional ranking in the early 1970s to almost bottom rank by the early 1980s. This reflects a dramatic shift in fortunes which is particularly pronounced for manual male earnings. For men, no other region saw earnings fall as sharply or consistently as the West Midlands. In the case of women, the East Midlands reflected a similar decline. On the other hand, both Scotland and the North, with relatively high unemployment, saw their earnings rise relatively, thus confounding any theory that labour excess alone lay behind the decline in relative earnings in the West Midlands. Within the region, higher pay rates were located within the West Midlands County, thus offering an incentive for some industrialists to locate elsewhere in the region (Smith 1984b).

Changes in the number of employees and their earnings in particular industries would affect average earnings in a region. In particular, the significant reduction in the relative earnings of motor vehicle employees in the West Midlands boosted, even if it did not cause, the fall in relative earnings for males. More generally, however, the slackening of demand for labour (reflected in unemployment) in the West Midlands seems to have eroded relative earnings. This occurred across both sexes and many industries. That this was the case to a greater extent within the West Midlands may, in part, be due to the very rapid rise in unemployment among some of the region's more highly paid skilled and semi-skilled workers.

The region moved down the regional earnings ranking in two sharp stages, falling six rankings in 1974–5 and a further three in 1979–80. The timing and scale of these falls matches closely the rise in relative unemployment in the region. However, if this is a factor influencing earnings in the region, it has not had the same effect in Scotland, the North, and the North West where unemployment and earnings have both been rising compared with other regions.

The decline in overtime working, the lack of increase in payment by results, a smaller increase in shift working, and persistently lower annual increases in earnings (excluding overtime etc.) occurred in the West Midlands to a greater extent than in other regions. The region's position in relation to shift work and payment by results implies less intensification of work in this region compared to others—notably Greater London, the South East, East Anglia, and the East Midlands. This has some significance for local labour costs. Thus supply and demand in the labour market provide a broad explanation for the fall in relative earnings in the West Midlands, while that fall was accentuated by the significant decline in the size of the motor vehicle work-force and the rapid decline in their relative earnings—reflecting malaise in that industry and its component suppliers rather than more general earnings erosion in the West Midlands economy (Smith 1984b).

The evidence indicates a close relationship to the pattern of increased unemployment, and such income reductions mean greater costs and pressures falling upon central and local governmental services (at a time of significant reduction in the resources available to local government). Local fiscal stress is the result.

Fiscal Stress

Fiscal stress can be defined as increasing demands on services not being matched by an increasing resource base. The metropolitan county council and its district authorities have experienced demographic and economic decline which has increased the pressures on local authority services. Declining population can reduce demand on services (for example, overall numbers to be educated), but it can also increase costs where higher income groups move out of an authority, which is left with an increasing proportion of lower income groups at the same time as its rateable value base is being eroded. However, the fiscal stress experienced by local authorities in the West Midlands County area stems not from a declining resource base but from a reduction in the level of government grant (Flynn 1983).

From 1977–8 to 1982–3 there was no significant decline in the rate base until 1982–3. Adding district and county grants together (at constant prices) for the West Midlands County local authorities, the figures show a real decline from £492.583 million in 1979–80 to £410.257 million in 1982–3. This represents a reduction in grant of 16.7 per cent or £82.3 million. The result has been a combination of price increases, rate increases, or cuts in services. Rates have increased but have failed to compensate for grant loss.

Cuts in expenditure which have resulted from loss of grant have had a negative multiplier effect on the area. In the case of Birmingham between 1981–2 and 1982–3 it is estimated that the loss of grant to the city and county councils has resulted in a switch of £23.1 million from direct private consumption. Overall, in Birmingham expenditure was increased less than was required to maintain services in real terms, with a significant loss of jobs. In many cases this reduction of expenditure and consequent job loss merely represents a transfer of resource expenditure from one government department to another—for example, from the Department of the Environment to the Department of Health and Social Security.

The central government block grant system by which local authorities receive the bulk of their government grants itself requires overhauling. There can be little forward planning by local authorities, while in Birmingham's case the situation arises where the city is only allowed to spend a budget which is less than that which the government's own calculations (Grant Related Expenditure Assessments) show to be required. If the city were to spend at this level, it would be penalized. In 1984–5 such penalties would be the highest of any Partnership area if Birmingham were to spend just above the spending target set by central government.

The Response of Large Firms to Decline

Our research concentrated upon twenty-six companies in the West Midlands with (in 1977 employment terms) the largest number of employees in manufacturing industry—the base which was being eroded. In 1977, just under 50 per cent of manufacturing employment in the West Midlands was in these twenty-six large companies, of which eleven had their headquarters in the West Midlands. Given the high degree of linkage in the local economy with many small and medium firms depending on markets generated by the large firms, then the influence of those companies extends well beyond their own work-force. They can be viewed as 'prime movers' in the local economy (Lloyd and Reave 1982). Hence we analysed the strategies and decisions of these companies within their international and national context. Their reactions to changing conditions are vitally important in determining outcomes in employment, production, and competitive position (Flynn and Taylor 1984).

Going beyond examination of indicators of economic performance since 1960, the important characteristics of the local economy included:

(*a*) A high level of industrial integration has been associated with the centrality of the vehicle industry (Dunnett 1980)

(*b*) Low unemployment and labour shortages in the 1950s and early 1960s made industrial relations difficult for management, and inefficient working practices reflected lack of control over the labour process
(*c*) Land scarcity and poor-quality premises constrained expansion and location decisions
(*d*) Regional policy grants and industrial development certificates (IDCs) altered the locational calculation for companies (Bentley and Mawson 1984, 1985a).

Beyond these local factors, other explanations for decline lay in national and international conditions (Mawson 1983c; Smith 1984e, 1985). Local factors could often be irrelevant to the main issues facing companies. Among these wider influences upon the activities of the top twenty-six companies were:

(*a*) Declining competitiveness with rising imports and poor export performance, especially significant in the vehicle and metal goods industries
(*b*) Erosion of Commonwealth preference and entry into the EEC necessitated a shift to a previously neglected market
(*c*) Domestic demand management and the use of the car industry as a regulatory tool altered the working environment of many firms
(*d*) The local economy was affected by national industrial policy, but this policy lacked coherence and interventions were made in the interests of the national economy, aimed at particular sectors or ailing companies, rather than with regard to local economic consequences (Bentley and Mawson 1984).

The recession changed operating conditions. High unemployment produced a fundamental shift in labour market conditions, allowing stronger managerial control of the labour process; the fall in aggregate demand severely increased competitive pressures; the expansion in consumer credit, despite the recession, created opportunities in consumer durable sectors; the banks launched a lifeboat of short-term credit for companies which might otherwise have faced liquidation; the value of sterling began to fall after a period of relatively high value.

The net effect of all this on employment was a dramatic reduction in these companies' work-forces. From 1972 to 1982, these twenty-six companies reduced their work-forces by almost 40 per cent. Employment at British Leyland in the county fell from over 84 000 in 1977 to an estimated 43 000 in 1982. GKN announced redundancies affecting over 50 per cent of its 1977 work-force, as did Lucas Industries. The largest firms were those which shed proportionately more labour, especially after 1978. Table 6.4 indicates employment loss by firm size for 1975–8 and 1978–81. It was those firms with over 500 employees which shed relatively most labour.

Such changes are the result of structural change that has been occurring since the mid-1960s, the present recession merely increasing the rate of

Table 6.4. *Employment Size Band Analysis for the West Midlands County Showing Net Change in Employment as a Percentage of Employment in the Base Year, 1975–1978 and 1978–1981*

Period	Employment size band					
	11–20	21–50	51–200	201–500	500+	Total
1975–8	−4.0	+3.5	−0.4	−4.7	−9.7	−5.7
1978–81	−24.7	−13.9	−19.5	−25.0	−32.9	−28.4

Source: Department of Industry, Regional Data System.

change. The adaptability of the local economy to diversify and prosper is restricted when the economy is dominated by a small number of large companies, whose strategies for survival may have beneficial effects for themselves, but adverse effects for the local economy.

Some firms began to diversify in the late 1950s and early 1960s. GKN moved from its metal manufacturing and metal engineering base towards a more heavy involvement in car components (especially transmission) and, in more recent years, has developed industrial distribution and services. Tube Investments moved from its metal manufacturing origins into consumer goods. Delta diversified into a broader spectrum of industrial activity. Others had to rationalize by closing new modern factories and plant—for example, Land Rover and Duport.

This diversification may be expected to benefit the wider local economy by switching from its narrow specialized base. However, the means by which the prime movers diversified—namely, acquisition—mitigated against such change. The common diversification pattern was to acquire or merge with another company which already had a stake in the new sector. Prime movers were shifting from the West Midlands, and acquisition generally took place outside the county area or region, sometimes in Europe to assist EEC market penetration. If it did not take place within the West Midlands, economic power was further concentrated. In other cases, firms switched their main manufacturing base overseas to relatively low-cost labour areas—for example, British Sound Reproducers moved to the Far East (Flynn and Taylor 1984).

With this shift by prime movers to other areas, the alternative route to diversification for the region lay through new firm formation or the attraction of firms or investment into the region. Regional policy increased the difficulties of the latter (though our work shows its impact was less significant than many have thought) (Bentley and Mawson 1985a), while the adverse image of the West Midlands as an ageing industrial area made it unattractive among possible sites for development. Thus new firm formation remained the only alternative. But here again, Fothergill and Gudgin (1979) suggest, the West Midlands scores badly.

Thus the methods adopted by large firms to deal with changing economic circumstances left a local economy still specialized in a narrow range of manufacturing industries, while the locus of production of the prime movers shifted from the West Midlands in their search for higher rates of return. The West Midlands was therefore left with an economy extremely vulnerable to recession.

Given the reduction in local markets, especially following the financial difficulties of British Leyland in the early 1970s, the components firms searched for new outlets. They exploited the expanding European car market. This had a profound effect upon the West Midlands, as not only were sales concentrated in Europe, but so also were manufacturing facilities. This and other shifts of production overseas were justified on various grounds by the large companies (Flynn and Taylor 1984):

(*a*) Lucas's first move into Europe was in 1959, initially to overcome trade barriers and give access to the protected EEC market

(*b*) To satisfy the nationalistic demands of governments and consumers the moves to Europe were often achieved by acquiring a company already operating, providing a base for expansion using existing contacts and the knowledge of local managers

(*c*) By diversifying the geographical distribution of assets (often beyond Europe), firms could maximize their potential for taking advantage of world patterns of demand and costs, and minimize the risk of losing assets owing to political and economic change in any one country.

Clearly the depth and severity of the recession took all the prime movers by surprise. Rationalization of production took place together with the large-scale run-down of capacity. Factories closed. Many were not old, obsolete, or inefficient. Some had been recently modernized, and it was precisely those debt payments on new investments which pushed some companies into crisis. The foundries scheme, encouraged by government aid, was one such cause of closures because of its timing (Bentley and Mawson 1984).

The critical question is whether the West Midlands economy is now smaller and weaker, caught in a spiral of decline, or is leaner, fitter, and better equipped to cope. The answer for prime movers may well be different from that for the local economy, or that for the work-force within that economy. Certainly a majority of the twenty-six firms are now reporting improved profits as they reap benefits from substantial improvements in productivity and performance. However, West Midlands industry now operates with a much smaller employment base. The structure of a number of large firms has also drastically altered. The outlook for those without work is not optimistic. Large sections of West Midlands industry no longer exist; competitive pressures are still intense, fuelling a continued impetus for further improvements in productivity and reductions in costs. Much of industry is operating under capacity. High interest rates and a high value of sterling by mid-1985 were themselves causes of considerable concern to manufacturers and made it extremely difficult to improve competitiveness. Even if demand were to dramatically increase, employment seems unlikely to rise. Thus a return to the low unemployment of the 1960s appears unlikely. The basic prosperity of the West Midlands as a growth area is no longer a realistic image of the area. It is an area struggling to adjust to a new economic role in the life of the nation.

Policy Response

A characteristic of the recent history of the West Midlands is the massive growth in the number of governmental (national and local) as well as other agency initiatives to review the problems of the local economy and unemployment. The notion of an 'experimenting state' well fits the policy response to the crisis of the local economy. However, these initiatives frequently have only marginal impact upon problems. The scale of the crisis is such that action by central and local government in conjunction with local business is called for. Local politicians are under pressure to be seen to be dealing with unemployment. Policies proliferate, usually in an uncoordinated fashion, and often with more concern for their initiation than their impact. Impact monitoring is rare. Organizational rivalries in this field are not uncommon while local authorities compete with each other in attracting industry. The West Midlands County area has performed very poorly in its attempts to attract new industry; hence much of industrial policy must rely upon maintaining and assisting the modernization and diversification of existing industry as well as attempting to develop jobs in other non-industrial sectors. The policy response has often been diffuse and late and too often unrelated to the real problems of the West Midlands economy. The evidence on regional planning suggests that it reflects a picture of too many lost opportunities, while pressures on maintaining Green Belt land often reflect local rivalries rather than a real concern for the economy as a whole (Bentley and Mawson 1985b). There is an in-built tendency towards conservation and stability when new challenges to boost the local economy are raised—for example, the Warwickshire coalfield, industrial expansion by the National Exhibition Centre. Policy assessment and organizational capacity require careful analysis, and many conclusions in such areas are inevitably based upon judgements about impacts both intended and unintended (see Spencer 1984).

Central government

In the provision of national aid for investment, the West Midlands has fared particularly badly compared with other regions. Although the region had obtained (by 1983) £54 million through Section B assistance of the 1972

Industry Act—more than most regions—this was only a very small proportion of the total governmental aid through that Act. Department of Trade and Industry (DTI) regional aid has to be seen alongside other forms of aid: national selective assistance; general and sector schemes; new technology and small-firms support; as well as direct assistance to specific companies, notably for the restructuring of British Leyland (Bentley and Mawson 1984). In the West Midlands, aid was small compared with that given to the Assisted Areas. There was also the negative impact of IDC policy until this was suspended in 1979. This operated vigorously from the mid-1960s to 1971–2. Thereafter, there were few refusals. In practice, IDC policy relaxed; but publicly the policy was seen as operative. The result was that many firms had to take over nearby dilapidated factories or squeeze more production from existing premises. It is likely that IDC policy prevented some large-scale projects being undertaken in the region—for example, the expansion of Chrysler in Coventry (see also Healey and Clark 1984). Other implications were: the absence of advanced factory building was an inhibiting factor; efficiency and productivity declined through continued use of old premises; companies left the West Midlands (and these mobile companies were often more growth oriented). IDC policy was damaging, but its effects have been overstated (Bentley and Mawson 1985a). When the policy was vigorously applied in the 1960s, there was a serious labour shortage in the West Midlands and the policy contributed to relieving overheating pressures (Smith 1972; Mawson and Smith 1980). Also, loss of employment through plant movement was always only a small proportion of total employment change.

There has been periodic Department of Trade and Industry direct intervention in West Midlands firms. The Industrial Reorganization Corporation's activities in the late 1960s were designed to improve efficiency and productivity in key sectors (Hague and Wilkinson 1983). This had direct consequences for key West-Midlands-based firms and had geographical consequences for the West Midlands. However, it is clear that the policy operated as a national programme with little account for local firms, employment, infrastructure, or regional economy implications. The National Enterprise Board's (NEB's) role in aiding British Leyland had very wide multiplier impacts upon sectors and individual firms and overall motor vehicle strategy (Dunnett 1980; Edwardes 1983; Central Policy Review Staff 1975). This has led to a restructured and more efficient assembly sector, but left the interlinked metal manufacturing and foundry sectors to 'sink or swim'. There has not been any strategy development of significance for the vehicle industry and its component suppliers, with the latter often being omitted from consideration. Industrial reliance on a regular single customer has left many firms vulnerable while lacking the managerial capacity to diversify. The very buoyancy of the regional economy until recently has led to the very policy neglect which in turn has fuelled decline.

Schemes such as Ferrous Foundry Scheme led to companies investing in modern plant and machinery in a sector where there was already overcapacity and led some firms into difficulties in attempting to service high loan debts (Bentley and Mawson 1984). Many other schemes were not regarded locally or nationally as particularly successful. Reacting to all this, and with a disposition to assist small firms and encourage new technology, the present government has maintained and expanded significantly forms of industrial assistance which can apply across all sectors. These are usually dependent upon 'take-up', requiring firms to be aware of entitlement and financially able to invest their element of contribution to aid. The West Midlands has fared badly—possibly reflecting a tendency for self-sufficiency, and lack of understanding and use of government-initiated assistance, while the years of prosperity bred a non-innovative management style (Edwardes 1983; Williams *et al.* 1983).

The Team for Innovation was established in the regional DTI office to encourage take-up of government aid schemes. However, firms still have to find 70 per cent of the cost (too high for many). Industrial technology is complex, and it requires very specific skills to match the technology to the individual firm—hence the local view that many schemes are not too relevant for West Midlands industry. Certainly, the sometimes traditional, rather backward, management of many firms (some still run as family firms) requires a wide range of support—for example, in marketing, production, new technology, finance, and business planning. Such support clearly needs to be tailored to the needs of individual firms. Policy needs to be formulated both centrally and locally against the background of clear business understanding and likely impact not just in general policy terms but in industry-specific terms. This may well mean that policy has to be flexible enough to cope with a wide variety of local circumstances.

The DTI has not adequately related national industrial policy to its spatial consequences. The conurbation or regional impacts of the actions of the Industrial Reorganization Corporation (IRC), the NEB, sector working parties, and innovation schemes have not been explored until recently, and then only in relation to the Team for Innovation. The DTI has a national perspective and the role of the regional office in the West Midlands has been primarily as an executive agency administering nationally developed schemes, though pressure on behalf of local perceptions has been applied by regional officers.

Since the early 1970s the regional office has argued the case that the West Midlands has had a structural problem, but this was not accepted until the 1980s. A fundamental reappraisal of regional policy nationally posed political problems as well as problems for government departments—that is, the new regional problem was being redefined as related to urban decline and high unemployment. Thus interdepartmental policy questions were raised, as were issues about the role of local authorities, which proved too difficult to face. The concept of a co-ordinated urban policy among government departments has proved too dificult to achieve. Perhaps our governmental structures require reappraisal, together with a fundamental review of the role of regional offices of government and how the latter might more effectively work jointly.

All this leads to a related issue. By the 1980s the Department of the Environment was in practice engaged in a territorial programme of economic development though maintaining a social bias—a trend, which together with some local authority economic development initiatives, had been contradicting the DTI's inter-regional policy. Urban programmes, land registers, enterprise zones, free ports, enterprise trust encouragement, Urban Development Grants (UDG's), management of local authority capital programmes, regional structure and local planning, and housing and transport investment strategies, all have direct regional economic impact.

The Department of the Environment has a greater and more significant regional presence than the DTI and manages considerable financial programmes regionally. It also has a higher grading of staff with a greater proportion of senior administrative posts, and a direct relationship with local authorities through a number of its policy initiatives. Both the DTI and the Department of the Environment have an involvement in economic development at the local level. The co-ordination mechanisms are vested in a regional economic planning board, but it meets infrequently. Officials liaise on an informal basis on projects and schemes where there is overlap. However, these mechanisms simply do not provide sufficient scope for policy debate and co-ordination in a strategic sense. Individual *ad hoc* programmes and their management dominate. The debates on regional and inner city policy need to be drawn together. Compartmentalism is not in the best interests of the welfare of the West Midlands economy. New mechanisms are needed at regional level which have a greater capacity to be responsive to local issues.

Department of Employment and Manpower Services Commission (MSC) schemes play a vital role in the market-place for school-leavers. In the West Midlands there is a very high probability that a school leaver will have as a first destination either a Youth Training Scheme (YTS) place or a job under the Young Workers Scheme (YWS). The MSC and Department of Employment are not so much intervening in the market-place for school-leavers as dominating it. The other traditional route into skills training, the apprenticeship, which used to be available to a small proportion of school-leavers, is already much restricted. Between 1979 and 1981 there was a decline of 79 per cent in first-year apprenticeships in metal and machinery including motor services, resulting in only 229 such apprenticeships in 1981 (City of Birmingham Careers Service 1982). The prospects for school-leavers are bleak.

There is not a major demand for training young workers from employers, and only 12 per cent of Mode A YTS schemes are run by employers. There are problems with private-agency-led schemes, difficulties in finding places with private employers, poor training quality, and discrimination in placing black school-leavers. There is no planning of the types of skills training to be offered and no attempt to match the skills produced by the scheme with the demand of local employers. There is almost a random collection of training offered. It is argued that the schemes produce transferable skills of numeracy, decision-making, and computer literacy. Yet this does not represent the sort of training for job opportunities which is now offered to adults (Flynn 1984).

The YTS, YWS, and Community Programmes (CPs) are run as separate schemes. There is a lack of a linked package from school onwards, which has caused particular concern among local education authorities. Many of those involved in the programmes are concerned about the post-YTS situation and the lack of an adequate training element on the CP. Placement systems have been viewed as racially discriminating in their allocative systems. The changed employment market has affected the motivation of some of those near the end of school-days. At the same time, there are questions of whether existing education is preparing people for the reality of life in the 1980s. The danger is that the skills needed for industry and commerce will not be there if and when required.

To many companies operating in the West Midlands, government purchasing policy is of equal, if not of greater, importance than any policy aimed directly at their industry. For example, the Coventry plants of GEC which manufacture telephone and telecommunications equipment are heavily dependent upon the purchasing programmes of British Telecom and will be affected by the liberalization of equipment following privatization. Defence spending is crucial to companies such as Lucas Aerospace and to parts of GKN. Such programmes are not usually considered in terms of regional effect but may, in terms of specific companies and the regional economy, be particularly significant.

The appointment of a Minister for the West Midlands in 1982 highlighted interdepartmental confusion and problems. Specifically, the Minister was (until the post ceased in late 1984) responsible for the Innovation Team, the English Estates Corporate initiative, the West Midlands Industrial Development Association, and all matters relating to DTI industrial policy. However, he was widely seen locally as the Minister responsible for all matters relating to the West Midlands economy. His appointment drew attention to problems of overlap and co-ordination. Inconsistency can lead to contradictions between policies within central government. For example, the Department of the Environment operates the UDG scheme which is a subsidy to encourage jobs on schemes which would not be developed without grant. The DTI does not fund projects on job creation grounds which are less than £0.5 million. Hence two regional offices are applying different criteria for the allocation of very similar grants to companies, while the DTI has only recently become involved in assisting Department of the Environment decisions on UDGs.

At the end of November 1984, following the review of regional development policy, it was announced that the whole of the West Midlands County area, together with the outlying towns of Warwick, Stratford-upon-Avon, Lichfield, Tamworth, and the New Towns of Telford and Redditch, was to be given Assisted Area status. Category 2 aid will be available (the inner core of Birmingham was not given Category 1 status despite an exceptionally strong case). These areas will qualify for selective assistance and for EEC Regional Development Fund resources (see Mawson *et al.* 1983). Given the planned overall budget reduction of £300 million in the regional aid programme by 1987–8, it might well be argued that this is really a case of too little too late. Nevertheless, it should result in some ameliorative effect on what would otherwise have become an even worse situation. The access to further EEC funding which such a decision incorporates is important. Case-studies show that authorities which organize effectively can secure assistance: much will depend upon the capacity of local authorities to develop these links and their advocacy role (Mawson *et al.* 1983).

In the West Midlands Region there exists a collection of policies and non-policy areas, which have had an impact on the regional economy, but government lacks any coherent view of the impact of policy. Ministries operate individual policies, each with its own aims and policy instruments. Local authorities pursue policies which may be contrary to those of central government in intent and approach, or may simply exist to fill a central government policy vacuum. The approach to policy issues is that new initiatives grow by accretion. There is not an analysis of the problems of the economy of the conurbation or the region as a whole, nor of the possible relevant interventions government policy might take. One reason is that regional offices of government are seen primarily, if not solely, as implementation arms of central government. The scope for policy analysis or strategic thought at regional or more localized levels is limited, especially as the research and intelligence arms of the regional offices have been greatly reduced in capacity in recent years. Combined with the abolition of the metropolitan county in 1986, this poses a very real and significant problem.

Without a proper analysis of the economic and social changes occurring in the local economy, there can be no coherent policy formulation nor effective priority determination. The present pattern of regional spending appears to be the outcome of a series of unconnected budget allocations by individual ministries in London. For example, in 1984, spending plans by the Department of the Environment on canals in the region exceeded what the DTI was planning to spend as a result of the Team for Innovation's efforts in promoting take-up of schemes to encourage industrial stimulation in the region. If policies are designed to assist the local economy, a greater concern with the alternative use of government resources in the local economy and a systematic monitoring and review of the effectiveness of policies is desperately required. There is too much concern with policy initiation and the launching of new *ad hoc* policies without a realistic assessment of policy impacts upon the West Midlands including those impacts designed specifically to assist and the consequential effects which flow from other policies designed with different purposes in mind (for example, trading policy, interest rates, domestic consumption controls, and exchange rate policy).

Local government

In the period between 1974 and 1979 an increasing awareness of the growing seriousness of the decline of the West Midlands economy led, particularly in the later 1970s, to the development of *ad hoc* and limited policy responses by local authorities. The West Midlands County Council highlighted the likely future problems of the economy in 1974 (West Midlands County Council 1974). This influenced key professionals but did not really affect local or national political agendas. Local authority responses were by individual service committees and departments, especially in the area of land and premises. There was limited organizational change, together with some development of an analytical capability on local economies (Mawson and Naylor 1985).

This period in general was characterized by the dominance of the Urban Programme initiative in the authorities affected by it, Birmingham and Wolverhampton. There was a recognized need to co-ordinate programmes between departments, but this was generally hindered by a lack of clarity about committee responsibilities in economic development and by little serious policy discussion. The Inner City Partnership scheme in Birmingham made the situation there more formal, a key objective of the partnership being economic regeneration. Local initiatives tended to be Urban Programme led, and mainly infrastructure oriented, especially on land and premises—for example, site development, small-factory-unit construction (ahead of the private sector recognition of need) (Mawson and Naylor 1985).

Between 1979 and 1981 economic decline intensified. Unemployment had more than tripled and derelict land increased significantly as major well-known industrial premises closed: by 1982 there were 4800 acres of derelict land in the county compared with 1000 acres in 1974. Disinvestment accelerated. During 1980 and 1981 the policy agenda shifted. The previous regional strategy with its complex organizational structures had been dominated by a geographical and environmental perspective on investment and production, with a focus on industrial overspill to expanding and new towns in the region.

By 1979 a new regional economic development strategy was well in preparation, advocating a programme led by sectoral investment, supported by an enhanced local authority role, and linked to a strategy of urban regeneration and containment of overspill. This policy shift was reflected in the West Midlands County Council's alteration to structure plan proposals (West Midlands County Council 1983). Local authorities and the regional Department of the Environment office, with local DTI office support, developed the proposals. Nationally, however, the DTI pulled out of the joint exercise, feeling there was not a serious regional problem. The pressures developed so that the traditional DTI regional perspective could no longer be maintained, and this led to the appointment of a Minister for the West Midlands in 1982 and to the announcement of regional aid for the West Midlands in November 1984. Meanwhile the Department of the Environment—via the extended Urban Programme, now covering Coventry and Sandwell in addition to Birmingham and Wolverhampton, and a series of new initiatives—had been developing its local economic programmes, albeit with a strong social content. In 1981 the Department of the Environment Enterprise Unit was set up locally to oversee the department's role in this field.

1979–84 saw the emergence of a series of *ad hoc* uncoordinated ministry-by-ministry programmes designed to respond to the crises of rising unemployment and economic decline. These presented both new foci and new funding opportunities for local economic and employment measures in local authorities. Some authorities were able to respond more than others, while local interdepartmental rivalries over declining local authority budgets and over responsibilities in the field of economic development were common. There was no coherent central government response to the problems facing the West Midlands. Within this vacuum, therefore, it is not surprising that local authorities began to develop their own responses as they saw fit (for details of these responses, see Mawson and Naylor 1985; Mawson 1983a,b,d).

The West Midlands County Council created its own Economic Development Committee, Economic Development Unit, and Enterprise Board in the 1980s (Mawson 1983a,d). It developed a set of policies based upon several key principles, including the need to strengthen traditional industries. A set of priorities were determined in 1981, largely using Section 137 funds of the 1972 Local Government Act. At this stage, little thought had been given to organizational arrangements and even less to the monitoring of policy effectiveness. The key issue was seen to be the initiation of new policies. Its early policies and programmes came under criticism. The County Council has funded firms and assisted them to remain viable, a few have had to cease trading, some have taken on more staff, and assets acquired in some firms have been sold at profit later. Its success has grown over the years, and the county has managed to develop a significant local economy monitoring role and to assist in helping local firms and creating and sustaining jobs in the local economy.

The county council has pressed through Partnership and Programme authority links for more people-oriented employment schemes, and for schemes tailored to specific firms—for example site acquisition and new premises. It has developed as the largest MSC training agency in the West Midlands, with a high placement ratio and schemes tailored for disadvantaged groups. It has encouraged co-operative formation, with forty new co-operatives and 200 jobs between 1981 and 1983 (at a cost of about £3000 per job) (Mawson 1983b). The role of welfare benefits has been highlighted, reflecting in 1984 a 50 per cent increase in persons on supplementary benefit in the county since 1981. Almost one in three households now live at or below the official poverty line. The county council's economy research and intelligence function has been very important, especially since the run-down of both the Department of the Environment and the DTI regional office capacities in this field. This research and intelligence capacity will be further reduced with the abolition of the Metropolitan County in 1986.

Between 1981 and 1984 the district councils within the West Midlands County developed economic development activities spurred partly by the increased access to Urban Programme funds, especially in Coventry and Sandwell (Mawson and Naylor 1985). Coventry adopted and expanded its limited economic programme, which remained infrastructure oriented. Birmingham established an Economic Development Committee in 1981, and much of the Partnership-led schemes remain concentrated on infrastructure, land assembly, and development, with a more recent emphasis on training schemes. Through main programme budgets, Birmingham has sustained a relatively large budget concentrating on land acquisition, recycling, and industrial refurbishment. Much of the activity is in the core area of the Birmingham Partnership zone where, in 1979, 37 per cent of the industrial buildings were built before 1914. The city has been prolific in new initiatives, bidding for UDGs liberally, and successfully (Wilde 1985), exploiting the National Exhibition Centre, and developing a science park. Birmingham reflects a greatly increased awareness of the nature of its economic decline and of its social consequences, but has only recently begun to consider the impact of its various initiatives and to develop a closer linking of initiatives to the problems of particular firms and to the problems facing particular communities and disadvantaged groups.

In the Black Country, Wolverhampton, Sandwell, and Walsall have been primarily Urban Programme led in their approaches to economic development, especially in the case of Sandwell where the authority was assisted by a three-person Department of the Environment Task Force for Smethwick during its first year of operation. This task force has since been disbanded at the request of Sandwell. Wolverhampton has its separate Economic Development Committee, but this has very limited terms of reference. All are affected by very limited budgets. Dudley is the site of an enlarged Enterprise Zone, which to date is one of the least successful zones nationally (Tym and Partners 1984). It also established its own Industrial Development Unit in 1978. Dudley and Wolverhampton had tended to be the more innovative among Black Country authorities. Other factors affecting Black Country initiatives and the capacity to mount them have been low rateable values, weak voluntary sectors, the sheer magnitude of the problem of derelict land concentration in the area, and the very small number of staff engaged in economic development activity work.

Table 6.5 sets out for each district council and the county council the pattern of expenditure within their 1983–4 budgets on local economic development initiatives. By this date, such initiatives had become wide-ranging (particularly within the county council and within Birmingham District Council). We have categorized expenditure under several key headings which summarize the types of initiatives being undertaken. These are industrial improvement areas, new and refurbished factory units, land provision, aid to industry, new enterprise workshops, special projects and new technology, projects to assist the unemployed, training, assistance to the economic circumstances of families, and industrial promotion. The relative spending patterns and policy emphasis can be gleaned from the data in Table 6.5. This demonstrates a much greater degree of sophistication in dealing with local economic problems, especially within the county and in Birmingham, than was the situation in the late 1970s and very early 1980s. Local authorities are adapting as they learn by doing and are beginning to rectify some of the omissions in their previous programmes. Nevertheless, resources are very limited and central government must continue to play a key role in funding such local economic development activity by local authorities.

There can be no doubt that local economic development initiatives are marginal to the scale of the problem. Nevertheless, they remain relevant to the problems faced. Equally there can be no doubt of the pressures upon local authorities to help their communities in easing the problems of economic decline.

Among the common problems facing local authorities in developing measures to combat economic decline have been problems of co-ordination between service departments and between government departments. The fact that European Social Fund and European Regional Development Funds criteria are frequently revised, and that they operate on a calendar year, has created problems (Mawson *et al.* 1983). Bids for Urban Programme funds are often rushed and oriented towards individual projects without clearly defined direction, while annual funding precludes longer-term planning (University of Aston 1985). The problem of slippage, particularly on economic development initiatives, means that often targets are not met and cash is spent on other projects (environmental or voluntary sector schemes). Local labour market statistics are regarded as inadequate by many having to plan training. The difficulties of incorporating an educational service perspective are common to most authorities, yet it is vital to developing an effective further education package relevant to local needs. Increasingly, authorities are having to piece together resources from different funding bodies and to develop a more entrepreneurial spirit. In some of the authorities, capacity to handle

Table 6.5. *Local Economic Development Budgets of Districts in West Midlands County, 1983–1984* (£000s)

District[a]	Industrial improvement areas	New and refurbished factory units	Land provision	Aid to industry	New enterprise workshops	Special projects and new technology	Projects to assist the unemployed	Training	Assistance to economic circumstances of families	Industrial promotion	Total
Birmingham:											
District	780	320	1 140	150	689	395	715	385	—	73	4 647
WMCC	887	676	590	60	—	—	1 123	1 086	105	19	4 546
Total	1 667	996	1 730	210	689	395	1 838	1 471	105	92	9 193
Coventry:											
District	—	36	—	—	352	33	35	—	—	—	456
WMCC	—	—	—	—	—	—	150	—	—	—	150
Total	—	36	—	—	352	33	185	—	—	—	606
Wolverhampton:											
District	450	4	667	35	50	—	293	76	—	105	1 680
WMCC	200	—	194	—	220	—	46	126	—	—	786
Total	650	4	851	35	270	—	339	202	—	105	2 466
Walsall:											
District	50	102	33	—	—	—	18	—	—	—	203
WMCC	50	—	15	—	—	—	—	—	—	—	65
Total	100	102	48	—	—	—	18	—	—	—	268
Sandwell:											
District	75	605	199	20	—	100	70	240	—	—	1 309[b]
WMCC	—	50	26	—	180	—	30	25	—	—	311
Total	75	655	225	20	180	100	100	265	—	—	1 620
District total	1 355	1 067	2 039	205	1 091	523	1 131	701	—	178	8 295[b]
WMCC total	1 137	726	825	60	400	—	1 349	1 237	105	19	5 858
Total expenditure	2 492	1 795	2 864	265	1 491	529	2 480	1 938	105	197	14 153[b]

[a] WMCC, West Midlands County Council.
[b] To these figures should be added a sum of £708 550 for economic schemes in Sandwell carried over from 1982–3.

elements of economic decline is weak. Where capacity is greater, both administratively and politically, the key stumbling-block often remains that of resources.

The Black Country authorities have co-operated in commissioning a joint study to assist economic rejuvenation. This has suggested a Black Country Development Agency. The establishment costs, however, are currently too considerable for the authorities to contemplate without governmental assistance. However, the exercise has indicated the distinctive nature of the problems of the local economies within the Black Country and has drawn the local authorities together to consider a joint response to their serious problems.

Abolition of the West Midlands County Council in 1986 leaves the aggregate ability of local authorities to combat economic decline greatly reduced. With the loss of the County Council's 2p rate (under Section 137), total resources will be greatly reduced. Districts will be more restrained by statutory main programmes and by real-terms declining resources, and will possibly be restricted to Urban Programme and other central-government-controlled initiatives as their means of dealing with economic decline. On the other hand, there remains some scope for the more imaginative use of main programmes, both by central and local government, to be directed towards alleviating the worst effects of economic decline. This is unlikely to happen on past evidence of inner city policy objectives. The prime reason for local innovative schemes within the county and district councils has been their access to resources (the 2p rate) without constraints and a varying response to take-up of various governmental schemes.

Conclusion

Local authorities in the West Midlands conurbation are struggling—some a little more successfully than others—with the new role of economic programme initiation and maintenance. Over time their capacities are improving through the learning process, but many established policies and procedures (Bentley and Mawson 1985b), as well as attitudes, are hindering progress. At the same time, it is central government that must play a crucial and more effective role in combating the problems of the conurbation and regional economies in the West Midlands.

At present, policy instruments available to local and central government tend to mould the policy prescriptions offered. It would be more satisfactory if policy formulation followed a process whereby a diagnosis of the problem led to the formulation of relevant policy interventions for which policy instruments could be designed and implemented by relevant agencies. Usually the opposite sequence occurs. This means that a closer tailoring of policies to match local needs is required if the West Midlands is to be positively assisted. Firmer targeting on relevant solutions to the area's economic problems is required; this will demand greater local understanding, and greater credence will have to be given nationally to the individuality of regions and local economies.

Is the new advent of selective assistance through regional policy likely to turn the fortune of the West Midlands? The Cambridge Economic Policy Group and our own analysis would suggest that even with the most favourable macro-economic policies and full Assisted Area status (not the present Selective Assistance status) the future of the region will remain rather bleak. If so, then there will need to be a greater concern than at present for the quality of life and the welfare of people in inner city areas which, together with some of the outlying council estates, are suffering the brunt of economic decline. That is where the reality of economic decline impinges upon individual well-being and, indeed, society's well-being.

All this raises much more fundamental issues about the nature of work and non-work, and about the pattern of rewards in our society—issues which have not begun to be realistically faced. The concept of two nations, one in poverty existing alongside the relative affluence of the other, is not a concept for the future. It currently exists and is manifest in many communities within, but not solely within, our inner city areas. Many policies and programmes fail, all too often, to reach those who are most in need of such aid. We need to be able to match aid more closely to the actual problems faced by intended recipients, whether they be firms, individuals, or geographical areas. A 1982 survey of the occupants of new and refurbished small industrial units in the Birmingham Inner City Partnership core area, which were funded by the county council with inner city resources, showed that in the fifty-eight companies only 16.8 per cent of the work-force lived in the inner city core area. Some 48 per cent lived in the much wider, complete, Partnership area which contained some two-thirds of Birmingham's population of one million (West Midlands County Council 1982). Where a large population is left without access to regular wage employment and depends on minimal state benefits, there is likely to be a growth in semi-formal and informal economic activity, such as low-wage personal services, street trading, and crime.

This is a period of rapid and increasing polarization not only in the allocation of rewards between rich and poor but also between sections of the labour force—between those with appropriate skills and those without, the employed and the unemployed, the regularly employed and the irregular and part-time workers. Polarization of the labour market has magnified the significance of the forms of social differentiation which affect entry to and promotion within employment—class, race, sex, and religion. Most clearly within the West Midlands, it is the members of ethnic minorities who will suffer disproportionately and for longer periods.

Any foreseeable adjustment of the economy will leave major problems to be addressed by social policy: these problems are different not only in scale but also in type from those which have hitherto confronted the welfare state. Our study demonstrates that, even if the West Midlands economy were to revive, without positive action by the public sector, it will leave a large proportion of the population in the miserable conditions associated with long-term unemployment and partial employment. A minimum programme for these groups requires at least three elements: first, a new concern with equity rather than the merest basic needs in income support, health, housing, and education; second, investment in public works and the caring services, not only in response to decay of the physical and social infrastructure of the cities, but also as revived sources of employment; third, a focus of the real job opportunities so created upon the young and long-term unemployed, with particular attention given to the needs of ethnic minority unemployment. Government has an important role here; so too do other public, private, and voluntary institutions.

The impact of de-industrialization has had its greatest national effect upon the West Midlands economy. Government and other bodies are slow to react to change, especially where traditional priorities and interests are strong and vocal. A vigorous programme designed to home in on the very real and serious issues facing the people of the West Midlands is essential and urgent. The alternative is a West Midlands left to continue to suffer economically and socially while a new geography of production, jobs, and social malaise is fostered in the 'national interest'. Political commitment to action and a radical rethink of present policies is necessary on the part of government and key institutions whose decisions have an impact upon the economic and social well-being of the West Midlands.

Note

1. West Midlands County Council, *Statistics 84*, 1985.

References

Ball, R.M. (1983), 'Spatial Structural Characteristics of Recent Unemployment Change: Some Policy Considerations', *Regional Studies*, 17(2), 135–40.

Batley R., (1984), 'A Historical Sketch of Industrial Change and its Social Impact in the West Midlands County', *Working Paper* No. 12, ESRC West Midlands Study, University of Birmingham.

Begg, I. and Eversley, D. (1986), 'Deprivation in the Inner City: Social Indicators from the 1981 Census', in V. A. Hausner (ed.) *Critical Issues in Urban Economic Development*, Vol. I, Oxford: Clarendon Press.

Bentley, G. and Mawson, J. (1984), 'Industrial Policy 1972–1983: Government Expenditure and Assistance to Industry in the West Midlands', *Working Paper* No. 6, ESRC West Midlands Study, University of Birmingham.

—— —— (1985a), 'The Industrial Development Certificate and the Decline of the

West Midlands—Much Ado about Nothing', *Working Paper* No. 15, ESRC West Midlands Study, University of Birmingham.

—— —— (1985b), 'The Economic Decline of the West Midlands and the Role of regional Planning—A Lost Opportunity', *Working Paper* No. 17, ESRC West Midlands Study, University of Birmingham.

Birmingham Chamber of Industry and Commerce (1984), *Regional Industrial Development: A Response to the White Paper*, Birmingham Chamber of Industry and Commerce.

Birmingham Inner City Partnership (1984), *Inner City Profile 1983–4*, Birmingham Inner City Partnership.

Central Policy Review Staff (1975), *The Future of the British Car Industry*, London: HMSO.

City of Birmingham (1984), *Land for Industry at April 1984*, City of Birmingham.

—— (1982), *Careers Service Report, 7/4/82*, City of Birmingham.

Cochrane, M.F. (1984), 'The Attitudes of Managers and their Wives towards Living and Working in Different parts of the UK', unpublished Ph.D. thesis, Centre for Urban and Regional Studies, University of Birmingham.

Compton, D. and Penketh, L. (1977), 'Industrial and Employment Change', in F. Joyce (ed.), *Metropolitan Development and Change*, Farnborough: Saxon House.

Confederation of British Industry (1983), *Manufacturing in the West Midlands: Problems and Prospects*, Birmingham: West Midlands Region CBI.

Danson, M.W., Lever, W.F., and Malcolm, J.F. (1980) , 'The Inner City Employment Problem in Great Britain, 1952–1976: a Shift–Share Approach', *Urban Studies*, 17, 193–210.

Department of the Environment, Inner Cities Directorate (1983), 'The Comparative Position of Inner City Partnership Areas', *Information Note* 3, Department of the Environment.

Dunnett, P. (1980), *The Decline of the British Motor Industry*, London: Croom Helm.

Edwardes, M. (1983), *Back from the Brink*, London: Collins.

Flynn, N. (1983), 'Fiscal Stress in the West Midlands County Area', *Internal Research Note*, ESRC West Midlands Study, University of Birmingham.

—— (1984), 'Manpower Services Commission and Unemployment', *Working Note* E, ESRC West Midlands Study, University of Birmingham.

—— and Taylor, A. (1984), 'Deindustrialisation and Corporate Change in the West Midlands', *Working Paper* 8, ESRC West Midlands Study, University of Birmingham.

Fothergill, S. and Gudgin, G. (1979), *Regional Employment Change: A Sub-regional Explanation*, Progress in Planning , Vol. 12, part 3, Oxford: Pergamon.

—— —— (1982), *Unequal Growth: Urban and Regional Employment Change in the UK*, London: Heinemann.

—— —— Kitson M. and Monk, S. (1984), 'Differences in the Profitability of the UK Manufacturing Sector between Conurbations and other Areas', *Scottish Journal of Political Economy*, 31, 72–91.

—— Kitson, M. and Monk, S. (1985), *Urban Industrial Change: The Causes of the Urban–Rural Contrast in Manufacturing Employment Trends*, Inner Cities Research Programme 11, Department of the Environment.

Gaffikin F. and Nickson, A. (1984), *Job Crisis and the Multi-nationals: The Case of the*

West Midlands, Birmingham Trade Union Resource Centre, Birmingham Trade Union Centre for World Development.

Green A.E. (1984), 'Considering Long-term Unemployment as a Criterion for Regional Policy Aid, *Area*, 16(3), 209–18.

Hague, D. and Wilkinson G. (1983), *The Industrial Reorganisation Corporation—An Experiment in Industrial Intervention*, London: Allen and Unwin.

Healey, M. and Clark, D. (1984), 'Industrial Decline in and Government Response in the West Midlands: The Case of Coventry, *Regional Studies*, 18(4), 303–18.

Johnson, A. (1982), 'Metropolitan Housing Policy and Strategic Planning in the West Midlands, *Town Planning Review*, 53(2), 179–99.

Karn, V. (1983), 'The Dynamics of the Urban Housing Market and its Impact on the Inner City—Processes of Change in the Birmingham Public and Private Housing Markets', *Working Note* F, ESRC West Midlands Study, University of Birmingham.

Keeble, D., Owens P.L., and Thompson, C. (1983), The Urban–Rural Manufacturing Shift in the European Community, *Urban Studies*, 20, 405–18.

Liggins, D. (1977), 'The Changing Role of the West Midlands Region in the National Economy', in F. Joyce (ed.), *Metropolitan Development and Change*, Farnborough: Saxon House.

Lloyd, P. and Reave, D. (1982), 'North-West England 1971–77: A Study in Industrial Decline and Economic Restructuring', *Regional Studies*, 16(5), 345–59.

Massey D. and Meegan, R. (1982), *The Anatomy of Job Loss: The How, Why and Where of Employment Decline*, London: Methuen.

Mawson J. (1983a), 'The West Midlands Enterprise Board', *Working Note* A, ESRC West Midlands Study, University of Birmingham.

—— (1983b), 'Co-operative Development Agencies: The West Midlands Experience', *Working Note* B, ESRC West Midlands Study, University of Birmingham.

—— (1983c), 'Explanations for the Decline of the West Midlands: A Summary of the Debate', *Working Note* C, ESRC West Midlands Study, University of Birmingham.

—— (1983d), 'The West Midlands County Council Local Economic Development Initiatives', *Working Note* D, ESRC West Midlands Study, University of Birmingham.

—— Gibney, J., and Miller, D. (1983), 'The West Midlands and the European Economic Community', *Working Paper* 4, ESRC West Midlands Study, University of Birmingham.

—— and Naylor, D. (1985), 'A Summary of Local Authority Economic Development in the West Midlands', *Working paper* 16, ESRC West Midlands Study, University of Birmingham.

—— and Smith B.M.D. (1980), 'British Regional and Industrial Policy during the Seventies: A Critical Review with Special Reference to the West Midlands in the 1980s', *CURS Working Paper* 72, Centre for Urban and Regional Studies, University of Birmingham.

—— and Taylor A. (1983), 'The West Midlands in Crisis: An Economic Profile', *Working Paper* 1, ESRC West Midlands Study, University of Birmingham.

Smith, B.M.D. (1972), 'The Administration of Industrial Overspill', *CURS Occasional Paper* 22, Centre for Urban and Regional Studies, University of Birmingham.

—— (1983), 'Population Change in the West Midlands', *Working Paper* 2, ESRC West Midlands Study, University of Birmingham.

—— (1984a), 'The Public/Private Sector Split in Employment in 1971 and 1978 in the West Midlands County: An Exploratory Study', *Working Paper* 7, ESRC West Midlands Study, University of Birmingham.

—— (1984b), 'The Labour Factor as an Explanation for Economic Decline in the West Midlands Region and County: Earnings in the West Midlands', *Working Paper* 9, ESRC West Midlands Study, University of Birmingham.

—— (1984c), 'Changes in the Number of Self-employed in the West Midlands Region', *Working Paper* 10, ESRC West Midlands Study, University of Birmingham.

—— (1984d), 'Employment Change in Parts of the West Midlands Region and the ESRC Inner Cities Project Study Area 1961–1981', *Working Paper* 11, ESRC West Midlands Study, University of Birmingham.

—— (1984e), 'Alternative Theories about Economic and Social Development, Decline and Change Focussing on those that Help to Explain Britain's and hence, the West Midlands' Decline from World Leadership in 1850 and particularly its Recent Loss of Relative Position', *Working Paper* 14, ESRC West Midlands Study, University of Birmingham.

—— (1985), 'Alternative Explanations for the Decline of the West Midlands', *Working Paper* 18, ESRC West Midlands Study, University of Birmingham.

Spencer, K.M. (1984), 'Evaluation of Public Sector Intervention and Approaches to the Concept of Institutional Capacity', *Working Paper* 13, ESRC West Midlands Study, University of Birmingham.

Stewart, J. D. (1981), 'The Dilemmas of Urban Public Finance', in R. Groves (ed.), *Economic and Social Change in the West Midlands*, CURS Occasional Paper 9, Centre for Urban and Regional Studies, University of Birmingham.

Taylor, A. (1983), 'Employment Change in the West Midlands', *Working Paper* 3, ESRC West Midlands Study, University of Birmingham.

—— (1984), 'The Changing Pattern of Unemployment in the West Midlands Region and County', *Working Paper* 5, ESRC West Midlands Study, University of Birmingham.

Tym, R and Partners (1984), *Monitoring Enterprise Zones, Year Three Report*, London: Department of the Environment.

University of Aston (1985), *Five Year Review of the Birmingham Inner City Partnership*, Executive Summary, Public Sector Management Research Unit, Management Centre, University of Aston, Birmingham.

West Midlands County Council (1974), *A Time for Action*, West Midlands County Council.

—— (1982), *New Small Units and Refurbishments: The Extent to which they Provide Employment for Inner City Residents*, Report to Economic Development Committee, 9 Nov., West Midlands County Council.

—— (1983), *Structure Plan: Proposals for Alteration, Technical Appendix*, West Midlands County Council.

West Midlands Economic Planning Council (1971), *The West Midlands: An Economic Appraisal*, London: HMSO.

West Midlands Forum of County Councils (1982), *The State of Housing in the West Midlands*, West Midlands Forum of County Councils.

—— (1985), *West Midlands Strategy Review. Regenerating the Region*, West Midlands Forum of County Councils.

Wilde, P. (1985), 'Urban Development Grants: The First Two Years', *Working Paper* 19, ESRC West Midlands Study, University of Birmingham.
Williams, K., Williams, J., and Thomas, D., (1983), *Why are the British Bad at Manufacturing?*, London.
Wood, P.A. (1976), *The West Midlands*, Newton Abbot: David and Charles.

INDEX